VOLCANOES OF THE UNITED STATES

ELLEN THRO

FRANKLIN WATTS
New York ■ London ■ Toronto ■ Sydney
A Venture Book

Photographs copyright ©: U.S. Geological Survey, Photographic Library: pp. 6, 9, 28 (D. H. Richter), 34 top (C. D. Miller), 45 top left (P. T. Hayes), 45 top right (L. Glover III), 45 bottom, 56 bottom (all H. T. Stearns), 69, 80, 98 (Roy A. Bailey); Grant Heilman Photography: pp. 31, 34 bottom (Alan Pitcairn), 47, 56 top; National Park Service: pp. 38 (M. Woodbridge Williams), 49 (Jack Boucher), 53 (Ralph Anderson); R. & D. Aitkenhead/ Positive Images: p. 43 top; Photo Researchers Inc.: pp. 43 bottom (Paolo Koch), 62 (D. L. Coe/National Audubon Society), 75 (Francis Gohier); Diane Kulpinski: p. 52; David K. Yamaguchi, Ph.D., Boulder, CO: p. 77; Dr. Michael Garcia, University of Hawaii: p. 87; UPI/Bettmann: p. 91.

Library of Congress Cataloging-in-Publication Data

Thro, Ellen.
Volcanoes of the United States / Ellen Thro.
p.   cm.—(A Venture book)
Includes bibliographical references and index.
Summary: Explores volcanoes and volcanic activity in the United States, discussing the study of volcanoes, their effect on the environment, and volcanic hazards and risks.
ISBN 0-531-12522-X
1. Volcanoes—United States—Juvenile literature.   2. Volcanism— United States—Juvenile literature.   [1. Volcanoes.]   I. Title.
QE524.T46   1992
551.2'1'0973—dc20   91-36002 CIP AC

CONTENTS

CHAPTER 1

VOLCANO COUNTRY, U.S.A.

"The most perfect mountain in the Pacific Northwest" is what many people called Washington state's Mount St. Helens, located 60 miles east of Portland, Oregon, in the Cascade Mountains. Mount St. Helens *was* beautiful in 1978. It tapered gently to an almost perfectly pointed snowcapped summit.

The mountain's beauty made people forget the forces deep within it, forget that Mount St. Helens is more than just a mountain. It is a *volcano,* a passageway to the surface for energy and materials from deep within the Earth. Its perfect summit was actually a crater, the site of an eruption almost 400 years ago.

Mount St. Helens hadn't erupted since 1857. By 1978, the mountain, along with its woods and lakes, was a favorite place for camping and hiking. Fishing was good in Spirit Lake, at the foot of its north face (or side) and in the nearby Toutle River. People who lived on the mountain called it a friend. Many didn't realize that the beautiful mountain was sending signals that its 121-year sleep was over.

After all, why should people worry? No volcano had erupted in the Cascades since California's Mount Lassen in 1914–21. Then in 1978, scientists at the United States Geological Survey published a report

Mount St. Helens before erupting

that said Mount St. Helens was the Cascades volcano most likely to erupt. They didn't know when, only that it could be during the next twenty years.

It didn't take twenty years. Only two.

Here's what happened.

Mid-March, 1980. Mount St. Helens experiences more than 170 progressively stronger earthquakes. Local residents scoff at scientists' warnings of possible danger. Then, over a three-day period, a tremendous explosion, or maybe two explosions, creates first one, then a second new crater inside the old one. The force sends a mile-high plume of volcanic *ash*—small pieces of volcanic rock—and steam into the sky. Some residents are evacuated. Eventually the two new craters become a single large one.

March 31. The earthquakes are joined by a special vibration pattern, a signal that the fluids and gases inside the mountain are moving. Ash and steam continue to appear. A few days later, the governor of Washington declares a state of emergency. She calls out the National Guard to block the roads, so no one can enter the danger area.

After several weeks, a large bulge appears on the north side of the mountain, at a place called Goat Rocks. It is caused by molten rock, or *magma*, moving underground like a burrowing mole.

May 12. The bulge is almost six stories high. As it continues to move, the surface behind it sinks.

May 18. Today is Sunday, and the first day of the trout fishing season. Eager campers are in the Mount St. Helens region but away from the evacuated area. In Spokane, some 200 miles to the east, people have gathered for the annual Lilac Festival. Volcano scientist David Johnston has been on duty all night at an observation post 5 miles north of Mount St. Helens. At 7:00 A.M., he reports by radio to the Volcano Observatory in Vancouver, Washington, that all activity is pretty much the same as it has been for the past several weeks. Scientists are also observing the mountain from airplanes and other ground stations. They, too, see nothing different.

At 8:32 A.M., a large earthquake occurs about a mile underground. Within 10 seconds, the Goat Rocks bulge collapses. David Johnston quickly radios, "Vancouver, Vancouver, this is it!"

During the next 15 seconds, a huge landslide moves down the mountain's north side at 150 miles an hour. The loss of this weight lets pressure underground burst through the surface. It is similar to what happens when firefighters remove the cap from a fire hydrant, but on a much bigger scale.

The Goat Rocks bulge blasts open in two places. Four hundred miles away, the sound is as loud as artillery fire. The explosion releases a torrent of water that was heated underground to 600°F, far above the boiling point. A mixture of rock, ice, soil, and ash shoots almost half a mile into the air and as far as 12 miles north. Rocks as big as 65 feet across are later found miles away. Trees are sheared off at the ground.

Most of the material falls back onto the mountain slope. The water turns to steam. In a few seconds, the steam mixes with the rocky particles to form a glowing cloud that turns the surface soil into mud. This muddy mass moves down the north side of the mountain, leveling everything in its path.

Meanwhile, another explosion has blown out the south side of the crater itself. It literally takes the top of the mountain off—the top 1,355 feet. A mushroom-shaped cloud of steam and ash billows 10 miles into the sky. Its electrically charged particles create spectacular lightning and a blue halo around the Sun. At midafternoon, Spokane is in total darkness from falling ash. Ash is more than an inch deep within 200 miles of the mountain. Overall, 1.7 billion tons of ash are deposited on ten states.

By evening, when the mountain quiets down,

(Top) The eruption of Mount St. Helens
(Bottom) Mount St. Helens after the eruption,
1 to 1½ cubic miles of material.
Notice the tremendous difference in
the shape of the mountain.

David Johnston and more than fifty other people have died. The mountain has emptied itself of 1 to 1½ cubic miles of material. A cubic mile is about eight city blocks long, wide, and high.

VOLCANIC AREAS OF THE UNITED STATES

The eruption of Mount St. Helens was 1980's flashiest display of volcanic might in the United States, but it wasn't the only one. Volcanic activity takes many forms. About 600 miles to the east, in Yellowstone National Park, Old Faithful Geyser was shooting its water into the air every day, to the delight of tourists. In Alaska, steam and gases were coming out of the ground at the Valley of Ten Thousand Smokes, just as they had for the past sixty years. In California, engineers were generating enough electricity for a large city by pumping hot steam from deep underground. Across the state, earth rumblings were disturbing a quiet tourist area called Long Valley.

Volcanic activity can be found today, or it could be found in 1980, in many parts of the western United States, including Alaska and Hawaii, as Figure 1 shows. Almost all U.S. volcanoes are west of the Rocky Mountains, and most are in or very close to the Pacific Ocean. Volcanoes are found in many parts of Alaska. The Hawaiian volcanoes rose out of the ocean and built the islands that make up the state, a process still under way. Our easternmost volcano is in northeastern New Mexico, about 80 miles from the Oklahoma state line.

In addition to areas of volcanic activity, the United States contains areas of volcanic evidence—unusual formations and rocks. In New Mexico and Idaho, for example, you can hike across rocky lava formations created by ancient volcanoes. In Oregon and Hawaii, you can see hollow tubes of hardened lava, some of them big enough to walk through. In Arizona and Cali-

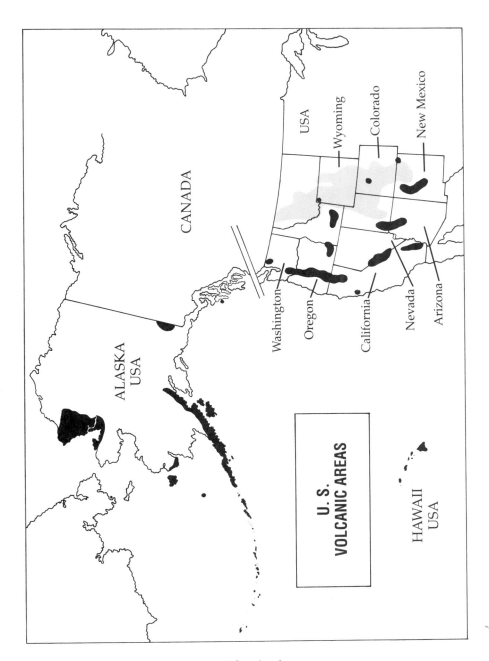

Figure 1. U.S. volcanic areas

fornia, huge shafts of ancient lava stand like oversized fence posts or lie shattered on the ground.

Even though today's volcanic activity is in the west, "Volcano Country" is bigger than that. The rest of the United States was part of it, too, at one time in the past. Both the Appalachian Mountains of the east coast and the Ozarks of the midwest had volcanic beginnings. In some places, ancient lava sits side by side with the products of much more recent activity. Elsewhere, the volcanic record is ancient and generally hidden far below the surface.

ACTIVE AND INACTIVE VOLCANIC AREAS

Many people think that lava-producing is the only kind of volcanic activity and that volcanoes are like light bulbs—they're either ON, sending lava down the mountainside, or they're OFF, totally quiet. Today, scientists use sensitive instruments to record eruptions and many other kinds of volcanic activity. A worldwide reporting network lets scientists know about eruptions or other activity that no one might have noticed earlier.

Modern studies have changed the way scientists think of volcanoes. A few decades ago, volcanoes were usually put into three groups: active, dormant (sleeping), or extinct (dead). Scientists still use these terms sometimes, but only for convenience, because they aren't accurate. Scientists have learned that sometimes it's hard to tell if a dead volcano is really cold or if a sleeping one might really be active.

Today a volcano is termed active if its last eruption can be given a recent date. This information may come from direct observation or written records, or it can be determined by technical analysis, such as carbon-14 dating. But volcanoes are much older than human history. To be on the safe side, scientists are now wary of all volcanoes active during the past 10,000 years.

Volcanism can also be active but not eruptive. It creates hot water geysers like Old Faithful at Yellowstone National Park, and wells and springs of hot water in other parts of the west. Several large underground sources of volcanically heated water and steam in California are being used to generate electricity.

For people who live in active areas, volcanism is a part of the hazards, risks, and rewards of everyday life. Even all these examples are just a part of the real "Volcano Country." The active regions of the continental United States are part of a worldwide volcano pattern. Volcanoes past and present have shaped the Earth's surface, affected its weather, and even played a big role in forming the air we breathe. Volcanism is also part of a basic behavior of our dynamic planet that goes back to its very origins, billions of years ago. In fact, everyone on Earth is a resident of "Volcano Country." The rest of this book explores our part of it.

OUR VOLCANIC EARTH

"The Ring of Fire"—The phrase hints at danger, mystery, maybe even something on a distant planet. It also describes the Pacific Rim, the Pacific Ocean shoreline that stretches from the tip of South America to the southern tip of Asia (Figure 2). The shoreline is fiery because it's lined with volcanoes, many of them active and eruptive.

When the solar system was forming, scientists think the Earth was a single molten, or melted, mass, the same inside and outside. Volcanic activity has been part of the Earth's life since the outer layer cooled into *crust* about 4.5 billion years ago. Gigantic volcanic eruptions, so big they are hard to imagine, occurred often. Studies of the crust show that almost two-thirds of it is *igneous,* or fire-formed, rock—rock that is volcanic or formed by volcanic heat. Volcanic gases brought metals to the surface, creating ore deposits. Over billions of years, volcanic activity built the continents.

It took 3 billion years (the Precambrian Era) to form the oldest part of North America, the Canadian Shield, including the Appalachian Mountains' deepest layers (Figure 3). The central part of the continent developed during the next 3 billion years, the Mesozoic Era. The

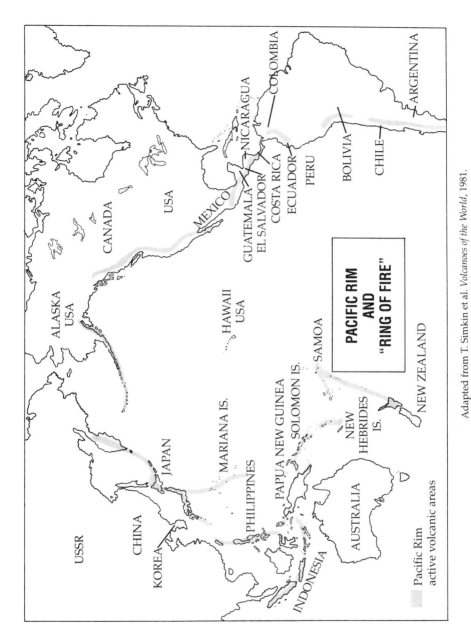

Adapted from T. Simkin et al. *Volcanoes of the World*, 1981.

Figure 2. The Pacific Rim and "The Ring of Fire"

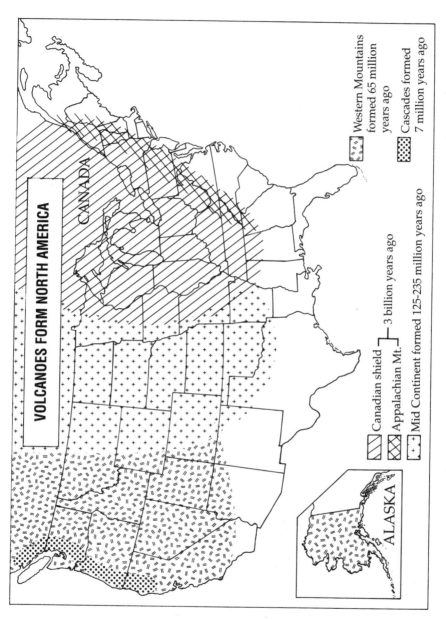

Figure 3. North America was
formed in great part
by volcanoes.

western mountains began forming about 65 million years ago. The oldest volcano craters in the Cascade Mountains are "only" about 7 million years old. (A *crater* is an opening in the Earth caused by a volcanic eruption or by the impact of a meteor.) Today's active volcanic areas are less than one million years old.

This is *what* happened. Scientists are still trying to solve the mystery of *how* and *why* it happened and to determine the origins of volcanic activity. They don't have all the answers, but they have the main ones. It turns out that the Earth is a huge heat-recycling machine that moves the continents from place to place like large puzzle pieces. Volcanoes are some of the machine's products.

THE GREAT HEAT RECYCLING MACHINE

Have you ever looked at a globe of the world and thought that you could fit South America right into the curve of Africa if only you could move them together? There's a reason they look like that. The two continents really were next to each other, millions of years ago. They've been moving apart slowly since then. In fact, all the continents, and the ocean floors, are moving. They are actually floating on another ocean, a layer of rock so hot (over 600°F) that it is really a very thick liquid, called magma.

Why is it so hot down below? First, a lot of heat is left over from the planet's formation. Second, some materials inside the Earth are radioactive and give off heat. Scientists estimate that at least half of the Earth's heat comes from radioactive decay.

Heat always moves from warmer to colder areas, whether it comes from your home furnace or the inside of the Earth. There are three ways this is done in the solar system. One is conduction through the crust. This is probably the main method on Mercury, Mars, and our moon.

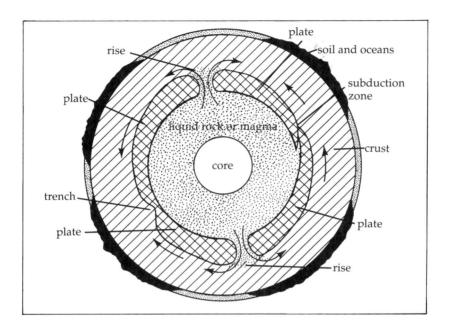

Figure 4. The structure of the Earth

The main method on Earth is recycling of hot interior material to the surface for cooling and then returning it down below. Scientists believe that recycling is responsible for removing two-thirds of the Earth's interior heat. It is the process that moves the continents, and the theory that explains it is called *plate tectonics.* Here's how our big heat recycling machine works. Refer to Figure 4 when reading the explanation below.

The Earth has a hot core of iron surrounded by a mantle of molten rock. Solid *plates*—the recycling machine's puzzle pieces—float and move on the mantle. Each plate has a base of dense rock, the *lithosphere* ("litho" means rock). The crust covers the lithosphere. The soil or ocean floor is the Earth's top layer. Each plate holds either an ocean floor or a combination of one or two continents and some surrounding floor. The

Earth's continents and oceans in relation to their plates are shown in Figure 5.

The United States is part of three of the Earth's plates (Figure 5). Most of it is on the North American Plate. This plate is manufactured by the Mid-Atlantic Ridge, which runs north-south. Hawaii and part of California are on the Pacific Plate, which is manufactured by the east-west Pacific-Antarctic Ridge. The small Juan de Fuca Plate (named after an early explorer) lies along the Pacific Northwest coast, between the two big plates. Some scientists think the southern end of the Juan de Fuca Plate is a separate, even smaller plate. They call it the Gorda Plate.

Magma moves upward at plate boundaries called rises (Figure 4). It cools and becomes new rock added to each plate. The plates move apart to make room for it.

Plates moving in different directions meet at boundaries called trenches. Either they scrape past each other, or one plate pushes the other down into the magma. The place where a plate is forced down is a subduction zone. The rock end of the plate forced into the magma is heated until it also becomes magma.

The Pacific and North American plates are slowly scraping past each other. Los Angeles and San Diego are on the Pacific Plate, while San Francisco is on the North American Plate. In about 12 million years, Los Angeles and San Francisco will be side by side. Juan de Fuca is moving east but is slowly being forced under the North American Plate. To the north, part of the Pacific Plate is being forced under the Alaskan part of the North American Plate. The most dramatic result of all this plate activity is the volcanic "Ring of Fire."

HEAT-RECYCLING VOLCANISM

The heat recycling system works just fine, but it's violent on the surface. Why? The moving plate bounda-

CONTINENTS AND PLATES

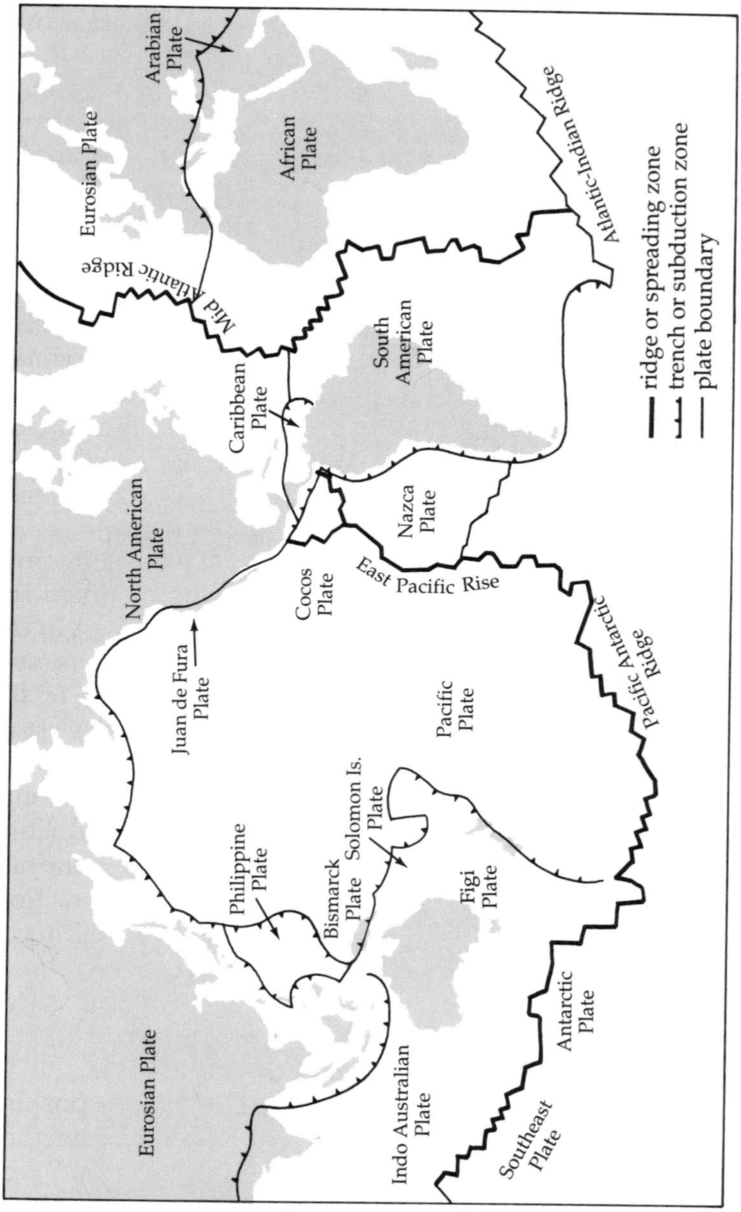

Figure 5. The continents and
the tectonic plates

ries are covered by rigid crust. When the plates move, the pressure and heat weaken the crust overhead and nearby. The crust stretches a little but finally must crack or change shape. The results can range from linear breaks, called *faults*, to entire mountain ranges. In California, the boundary between the Pacific and North American plates is along the San Andreas Fault. The 1989 San Francisco-area earthquake resulted from a disturbance along part of the fault.

Crustal breaks aren't clean cut, like cake slices. A broad zone of crust on either side of a boundary may form a network of faults or wrinkle up dramatically as a mountain chain. Magma underneath that is energetic can easily force its way toward the surface through the weakened crust, like water pushing through a weakened dam (Figure 6).

All volcanic activity, ancient or recent, results from magma interacting with surface materials. When the magma forces solid materials through the crust to the surface, an eruption occurs. The eruption may be violent or gentle, depending on the composition of the magma and the crust's structure and the degree of faulting.

The Cascade, Sierra Nevada, and Alaskan mountain ranges are all in the boundary and subduction zones of the three tectonic plates. They are examples of crustal weakening caused by the plates' interaction, and they are all highly volcanic. The entire "Ring of Fire" exists at similar plate boundaries.

OTHER CAUSES OF VOLCANIC ACTIVITY

Not all volcanic activity is caused by plates pushing against each other. The Earth has still a third method of releasing heat from the interior—"hot spot" volcanic activity, which throws heated material up through cracks in the crust at places that aren't plate boundaries. The Earth shares this method with only

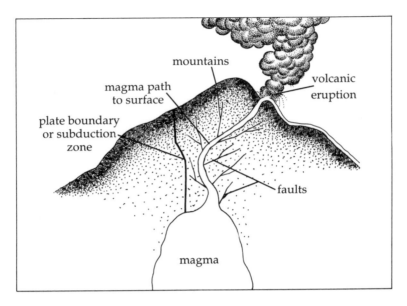

Figure 6. Faulting and volcanism.
Magma can move near or to the
surface through faults in the crust.

one or two places in the solar system—Jupiter's moon Io and perhaps Neptune's moon Triton. Hawaii's volcanoes probably result from especially hot active magma (a "hot spot") moving upward and stripping away the bottom of the moving plate. The magma then breaks through the weakened plate and crust. As the plate moves past the hot spot, the magma creates one volcano after another.

Some midcontinent volcanic activity may be at a zipper, a place where ancient continents came together and joined. The cause of other U.S. volcanic activity is still a mystery.

TYPES OF VOLCANIC ACTIVITY

There are several ways to classify the strength of volcanic activity. Scientists don't always agree, and the systems can overlap. Nonexplosive eruptions are often described as "Hawaiian" or "Icelandic." Moderate and large eruptions may be called "Strombolian" or "Vulcanian," after two Italian volcanoes. The largest are "Plinian" or "Ultra-Plinian," named after an ancient Roman who died during an eruption of Italy's Mount Vesuvius in A.D. 79. Other names for very violent eruptions are "Krakatoan," after an Indonesian volcano whose 1883 eruption had worldwide effects, and "Pelean," after a West Indies volcano, Mount Pelée.

Magma that flows onto the surface of the Earth is called *lava*. However, it doesn't always make it that far. At Mount St. Helens in 1980, for instance, the magma's pressure and heat created explosions of rock, steam, ash, and ice. The magma itself stayed below.

After an eruption is over, materials deposited on the surface may settle, become packed down, or collapse into the hollow beneath them. This leaves a crater or hole shaped like a basin, disk, kettle, cup, or tall glass.

There are two main volcano types:

strato volcano—which erupts from a central crater and looks the way most people think a volcano should look (Figure 7a); Mount St. Helens is an example.

shield volcano, with gently sloping layers of lava built up from many eruptions from many different craters (Figure 7b)—the type found in Hawaii.

Sometimes magma's heat moves through rock to an underground water supply. This heated water may stay in a well, or it may flow to the surface, becoming a hot spring. Depending on underground conditions, the water's movement to the surface may be explosive. An eruption of this hot water is known as a *geyser*, of which Old Faithful is one example.

Active magma contains sulfur dioxide and other gases, which it gives off as it begins to erupt or as it cools. The gases may filter out naturally as the magma

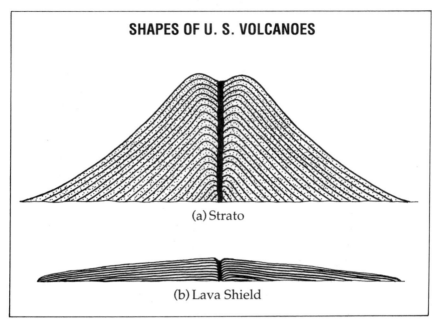

SHAPES OF U. S. VOLCANOES

(a) Strato

(b) Lava Shield

**Figure 7. The two types of
volcanic craters**

moves. If water is present, the gas may mix with it and make its way to the surface through cracks in the crust. Once on the surface, the gas may rise from the water. Gas, steam, or a combination emitted in this manner is called a *fumarole*, an Italian word meaning "giving off fumes." If the gas contains enough sulfur to be very smelly, it is called a *solfatara*, from the Italian word for sulfur. Alaska's Valley of Ten Thousand Smokes, for instance, gives off both steam and gases.

Volcanic activity is the sometimes violent product of an elaborate worldwide process, one that got started when the Earth was young and originates deep within it. Even so, the volcanoes and their settings are interesting for their own sakes, whether they're in Hawaii, the Pacific Northwest, Alaska, or elsewhere in the United States.

CHAPTER 3

ACTIVE
U.S. VOLCANOES

Lava flows. Rock blasts. Ash and steam. Earthquakes. Ground movements. The United States's active volcanoes are very active!

During the 1980s, twenty of them let people know they were alive and kicking. These are marked on the map in Figure 8. The most famous and powerful is Mount St. Helens. After its big blast, it erupted five more times that same year. If there were a volcano marathon, though, the winner would be Hawaii's Kilauea. It erupted in 1981 and again in 1982. Its next eruption began in 1983 and was still going on in late 1990, sending out quiet but relentless rivers and moving walls of lava. Kilauea's neighbor, Mauna Loa, erupted in 1984. In Alaska, Redoubt volcano made the evening news in 1989 and 1990 when it sent ash clouds over the city of Anchorage.

The others are less well known. Two, in thinly populated northern California, didn't erupt in the 1980s, but their earthquakes and ground surface changes told scientists that magma underground is moving. Another is a seamount or undersea mountain off the shore of Hawaii Island. Its earthquake record shows that it, too, has active magma. The other fourteen active volcanoes are in Alaska, ten in the remote Aleutian Is-

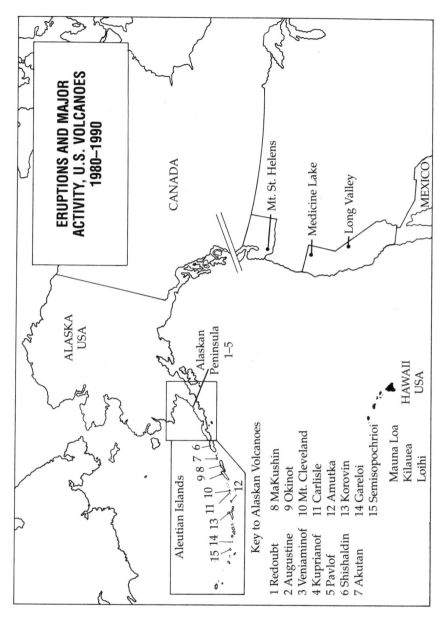

Figure 8. Eruptions and major
activity of U.S. volcanoes
1980–1990

lands. Eight of the fourteen actually erupted during the 1980s; the others produced ash or steam.

Most U.S. volcanoes are located in three areas: the islands of Hawaii, the Cascade Mountains (running through Washington, Oregon, and California), and Alaska. Of all the active volcanoes, those in Hawaii may be the most dramatic, because their roles in land building can be seen so clearly.

HAWAIIAN VOLCANOES

Almost ten years to the day after the major eruption of Mount St. Helens, another volcano reached out and slowly incinerated much of the town of Kalapana, on Hawaii's "Big Island," also named Hawaii. The volcano's name is Kilauea, and the eruption was still going strong in 1990, eight years after it had begun. It was the latest chapter in the volcano's life history, which began several million years ago.

The Hawaiian island chain stretches 1,500 miles from northwest to southeast. Each of the many islands is the product of one or more volcanoes. The direction also traces the islands' volcanic lifeline. Activity increases, and the islands are younger, the farther southeast they are. The western islands are no longer active, but volcanoes on all the other main islands of Oahu (Honolulu's home), Maui, and Kauai have had sustained eruptions during the past 10,000 years.

Hawaii Island, the farthest east, is the youngest and most active. Of the five volcanoes that formed it, four are active: Hualalai, Mauna Kea, Mauna Loa, and the youngest, Kilauea. If you stand on the beach at Hilo, the county seat on Hawaii Island, and look *mauka*, the Hawaiian word for "toward the mountains," you can see or imagine Kilauea's summit at a height of 4,000 feet. Towering even higher are Mauna Kea at 13,800 feet and Mauna Loa, 13,700 feet.

Or rather, these are the volcanoes' heights above

Hawaii's Kilauea, a shield volcano that
has been erupting since 1983

where you are standing. Actually, Hilo, at sea level, is more than halfway up the mountains. Their bases are on the ocean floor, more than 20,000 feet *below* sea level, making Mauna Loa and Mauna Kea the tallest mountains on Earth.

Several million years ago, there was no Hawaii Island, just a piece of ocean floor sitting over a magma "hot spot." When the magma's pressure split open the crust, a series of eruptions began. The earliest ones deposited lava on the ocean floor and left a crater. Over time, volcanoes grew as layers of erupted lava built gentle slopes, the trademark of what are called shield volcanoes. Each new volcano nestled on the south side of the next older one.

Finally, each mountain grew high enough to burst through the ocean's surface, with Kilauea the most recent. Air and sea erosion changed the shoreline. Without the weight of the water holding them down, the eruptions became more explosive. The lava changed, too. Water pressure made the undersea lava very dense and kept its gases from escaping. Above water, the gases were able to expand, making the lava lighter in texture. It was also filled with holes where the gases burst through.

Weathering and rainfall turned lava and ash into soil. Windborne seeds landed and sprouted. Birds and other animal life found homes there, and people from the South Pacific settled the land.

Periodically Kilauea erupted, creating one crater after another. Its gentle eruptions sent fresh lava to cover the vegetation and sometimes the towns. In this way, floods of lava provide new soil surfaces for Hawaii Island the way river floods wash new soil onto the land in other parts of the country. The island is growing outward, too. A new volcano, Loihi, is building under water against Kilauea's southern side.

Over time, Kilauea will become less and less active. Finally it will become extinct or die. As the centuries pass, air and water will wear it down to sea level. Coral reefs will grow around it under water. Finally, it will be an atoll—perhaps like the one shown in the photograph on page 31—a flat curve of land and reef around a lagoon. Right now, though, Kilauea is active. Some of its vibrations have shaken the entire island and set off large ocean waves called tsunamis.

Hawaii Volcanoes National Park is the home of Kilauea and the other active volcanoes. The others are much less active, however. Mauna Loa last erupted in 1984. Hualalai last erupted in 1800, and Mauna Kea 3,600 years ago.

Visiting the park is always an adventure, because parts of it may be off limits if the lava is flowing. Or you may have to take a long detour around barricaded roads. In 1989, the lava even destroyed one of the visitor centers.

Kilauea's crater is the source of some gases. But the lava and more gases come from cracks or vents farther down the slope. A vent called Kupaianaha is the source of the lava.

HAWAIIAN BELIEFS ABOUT VOLCANOES

The Hawaiian islands are entirely volcanic, so it's very natural that in the traditional Hawaiian religion the ruler of volcanoes is an important and powerful goddess. Her name is Pele (PAY-lay)—people call her Madame Pele—and her home is the volcanoes of Hawaii Island. Like the Hawaiian people themselves, Pele came from the South Pacific. Looking for a place to settle, she dug holes in the ground with Paoa, her magic stick. Each hole became a volcano.

The island of Oahu, where Honolulu is, hasn't had a volcanic eruption in a long time. Here, people say, Pele's holes filled with water, putting out the volcanic fires.

Pele has a bad temper. It is said that when she is angry, she stamps her foot, making the ground shake

An atoll, a flat curve of land
and reef around a lagoon

and starting an eruption. In the old days, Pele just threw her enemies into volcanoes or buried them in lava flows. This usually calmed her down. People could also calm her with gifts—usually fruits, but sometimes live animals.

Throughout the state, volcanoes and lava formations are connected to Pele legends. Lava thrown into the air sometimes forms small glass balls, called Pele's tears. Lava that forms long threads is called Pele's hair.

The surfaces of big lava rocks found along the shore often contain tiny rocks. In the Hawaiian language, the tiny rocks are called hanau, meaning "birth stones"—babies about to be born to the big rocks. Hanau are actually grains of sand washed by the ocean into weathered holes in the lava. They later hardened into rocklike material.

Even today, many Hawaiians think of volcanoes in terms of Pele and bring her gifts of fruit. Most people who live in the path of Kilauea's lava believe that the land belongs to Madame Pele. Even some of those whose houses have burned say she can reclaim her land whenever she wants.

CASCADES VOLCANOES

In the twentieth century, there have been just two volcanic eruptions in the lower forty-eight United States. Both were in the Cascades Mountains. One was Mount St. Helens. The other was Mount Lassen, in northern California, which erupted in 1914–21.

Both are strato volcanoes. This means they have erupted many times from a central vent. The volcanic materials, either lava or *pyroclastic flow* (a mixture of fiery pieces of rock, air, and volcanic gases), go downhill in all directions. As a result, the mountain's shape is symmetrical. The volcanic material builds up layer after layer. Strato (meaning "layered") volcanoes look the way most people think a volcano should look—tall and pointed, as Mount St. Helens was before 1980.

The Cascade Mountains could easily be called the Volcanoes Mountains instead. From Mount Garibaldi in British Columbia to Mount Lassen, the range contains more than thirty active volcanoes and volcanic areas. Most are within national parks or wilderness areas featuring beautiful scenery, hiking trails, and often snow sports.

The Native Americans in the region greatly respected the power of all the Cascades volcanoes. They have many legends to explain the mountains' rumblings and eruptions. These legends often tell of the volcanoes' power over people, and the death and destruction an eruption can cause. It is said, for example, that the mountains were once a group of men and women, forced to stop on a long journey. Some began to perspire, creating rivers. Some remained standing, while others sat or lay down. They stayed that way so long that they turned into mountains.

In another story, the land originally was flat. People who lived along the coast wanted their land to be fertile, so they tricked the Clouds and the Rain into giving them lots of water. But the Great Spirit thought the people were too greedy. To punish them, the Spirit scooped out earth, creating Puget Sound (in western Washington), then mounded it up to make the Cascades. The mountains blocked the movement of rain to the east. This is why the eastern portions of Oregon and Washington are deserts.

MOUNT SHASTA AND MOUNT MAZAMA

The Modoc Indians of northern California had a friendly explanation of one nearby volcano, Mount Shasta. To them, Shasta was a huge lodge with a central smoke hole, the home of the God of the Sky Spirits. Smoke and ash were simply signs that the god's fire

(Top) Mount Shasta, in northern California
(Bottom) Crater Lake is the lake in the crater of the collapsed
Mount Mazama in southern Oregon. The island is Wizard Island.

was burning. Scientists have been able to date nineteen of its eruptions, the last one in 1786.

Sometimes the mountains produced more than smoke and ash. Mount Mazama is in southern Oregon, just north of Mount Shasta. Mazama used to be 12,000 feet high, but about 6,600 years ago it had a tremendous explosive eruption. Ash fell as far as Saskatchewan, Canada, and covered more than half a million square miles. The explosion emptied the interior, and just an outer shell remained. The collapsed mountain left a hole half a mile deep. Rain and snow filled the hole, making what we now know as Crater Lake.

Before the lake filled, a smaller eruption created a slender, perfectly formed *cinder cone,* a cone-shaped mountain made of small chunks of volcanic rock or cinders. It even has its own crater. The cinder cone now rises from the lake and is called Wizard Island.

The Klamath Indians have a religious story that explains the great eruption. The god of the underground, Llao, lived on Mount Mazama. The god of the surface, Skell, lived on Mount Shasta. The two gods fought, throwing rocks and fire at each other. Llao's mountain collapsed, dragging him into the underground world, never to be seen again. Rain filled the hole in the mountain, and the hole became Crater Lake. In another version, both gods lived on Mount Mazama. Llao's head became Wizard Island. The tears his followers shed when he died became Crater Lake. The activity described in the stories is so accurate that scientists think the people who composed them actually saw the eruption.

Today Mount Mazama is the center of Crater Lake National Park. Visitors can drive to the volcano's rim and take boat trips around the lake or to Wizard Island. If you're athletic, you can climb a steep half-mile trail to the Wizard Island crater, which is about 300 feet across and 90 feet deep.

Other reminders of eruptions are all around. A large chunk of lava is known as Llao Rock. The remains of pyroclastic flow have left pointed spires called the Pinnacles. Merriam Cone, the evidence of another eruption, is under water.

MOUNT ST. HELENS TODAY

No tour of the Cascades would be complete without visiting Mount St. Helens. Now a part of Mount St. Helens National Volcanic Monument, the volcano has continued to be active. A series of small eruptions has built a lava dome—a mound of lava too thick to flow away—900 feet high and more than a half-mile wide on the crater floor. Vents alongside produce columns of steam. In recent years, underground activity has increased, but not enough to signal another major eruption. The mountain's slopes are still covered by a layer of ash, especially near the sharp and jagged rim, which is above the timberline. A coating of snow tops the ash for much of the year. But the mountain features life as well as signs of possible future destruction. Trees now grow and flowers bloom in the devastated area. Another type of life has returned to the mountain, too—people.

Unlike the older volcanic parks, there are no tourist accommodations at Mount St. Helens. People who want to go to the top must hike and climb the last 2½ miles, usually wearing equipment for climbing on snow, ice, and rock. Just as important, they must be prepared for the dangers of rain, fog, and windstorms at any time.

Despite these hardships, or maybe because the challenge equals the excitement of the volcanic activity, Mount St. Helens is more popular each year. In 1989, over 15,000 people climbed it, making it the second most climbed mountain in the world, after Japan's

Mount Fuji, also a strato volcano. In fact, people who want to climb Mount St. Helens must register with the U.S. Forest Service, which operates the monument.

There's an exciting way to make the return trip down the mountain if enough snow is on the ground. You can just sit down and slide for several miles— what climbers call a glissade (the French word for "slide").

ALASKA'S VOLCANOES

The Aleutian Islands, stretching toward the Soviet Union, and the rest of Alaska's Pacific coastline form part of the Ring of Fire. It's no surprise, then, that the state of Alaska has many active volcanoes, eighty-five in all. In fact, the Aleutians' forty volcanoes are among the most active in the entire United States. Their names reflect the region's location and history. Some names are Native American (Aniakchak, Okmok, Amutka). Some are Russian (Semisopochnoi, Pogromni). Still others are English, such as Frosty, Redoubt (meaning a barrier), Mount Gilbert, and Roundtop.

Many Alaskan volcanoes are far from places where large numbers of people live or work. They are often shrouded in fog or other bad weather, so they are hard to study on a regular basis. Most information about their activity comes from their few neighbors, from airline pilots, and from satellite pictures.

One group of volcanoes is relatively easy to get to and study. These volcanoes are on the Alaskan Peninsula, a bridge between the Aleutian Islands and the mainland of Alaska. Located south of Anchorage and the Cook Inlet, the peninsula contains thirty-two volcanoes. Several volcanoes are preserved in national monuments. Aniakchak National Monument is named for a volcano that erupted in 1931 and perhaps again in 1942. Its saucerlike crater, or *caldera*, contains cinder

Aniakchak National Monument in Alaska has a crater containing cinder cones, old lava flows, and a lake.

cones, old lava flows, and even a lake. Surprise Lake is so named because it plunges through a crack in the crater wall.

KATMAI, REDOUBT, AND THE VALLEY
OF TEN THOUSAND SMOKES

Katmai National Park, about 250 miles southwest of Anchorage, is a natural wonderland. It faces Kodiak Island (home of Kodiak bears) across the Shelikof Strait, a narrow arm of the ocean. Fjords (even narrower inlets), beaches, and coves line the shore. A system of rivers and ponds, left over from past glaciers, cuts through the landscape. At sea level, Katmai is filled with birch and spruce trees. Higher up, the land is tundra, a kind of Arctic prairie, covered with lichens and mosses. But Katmai has something more—volcanoes, many of them. Redoubt is there. The park is even named for a volcano.

In the nineteenth and early twentieth centuries, people believed Katmai mountain was extinct. Even so, it sat in a large field where vents gave off smoke and ash. But Katmai was just napping. In June 1912, it woke up spectacularly. First it announced that an eruption was on its way. Alaska has daylight almost around the clock in June, but Katmai produced enough smoke and ash to block out the sun. People had to breathe through wet cloths. Then the volcano erupted, producing large amounts of pyroclastic flow. The activity continued for almost two months.

Katmai erupted again in 1914 and possibly four more times, the most recent in 1931. The volcanic materials that fill the valley stayed hot enough to give off steam in many places, so many that people named it the Valley of the Ten Thousand Smokes. Sixty years later, the smokes continue, but there are far fewer than 10,000 of them (if there ever were that many).

In this book, it isn't possible to describe every active U.S. volcanic area, but you may wish to learn more about them on your own. The references at the end of the book can help you get started.

EXPLORING OUR VOLCANIC PAST

Volcanic activity has created or shaped the land we live on. There's a good chance that sometime soon you'll be walking or riding across a volcanic formation, probably without even knowing it. There are also many unusual reminders of past volcanism that people can explore and enjoy. These may be near or within active areas.

As Hawaii's story showed, after a volcano goes out of business and no longer builds or rebuilds, it is worn down by erosion and other natural processes. People may build homes and even cities on it. Eventually it may totally disappear as a landmark, or it will exist only in a few strange-looking formations. Not knowing the origin or nature of these formations, people often give them imaginative names, like Craters of the Moon or the Devil's Postpile. If you know what to look for, you can recognize the formations' volcanic origins. The easiest formations to distinguish are the ones that still have a "volcanic" shape, a mountain with a crater.

EXTINCT VOLCANOES

Volcanoes become extinct when the supply of magma runs out. The magma may move elsewhere, or as in

Hawaii, the plate may move elsewhere. In natural areas, the extinct volcanoes may look much like the active ones, at least on the outside. The Sisters is a group of three similar-looking mountains in the southern Cascades. Two are actively volcanic—North Sister and South Sister. Middle Sister is extinct.

In a city, an extinct volcano may be harder to recognize, especially if you don't expect to find a volcano downtown. People who visit Honolulu, Hawaii, for instance, don't always realize that some of the most famous landmarks in town are extinct volcanoes.

Oahu, the island where Honolulu is located, was once as volcanically active as Hawaii Island is today. Long ago, its volcanoes were built by lava flows, just as Kilauea is being built now. But, like Pele's water-filled stick holes, all but one of Oahu's volcanic fires are extinguished. Its old volcanoes have new identities.

Diamond Head (Leahi, to native Hawaiians) used to be the landmark for sailing ships approaching Pearl Harbor from Asia or the West Coast. It got its English name from nineteenth-century sailors who thought the glassy volcanic rocks were diamonds. Today Diamond Head's shape rising dramatically from the shore is still a striking sight to people flying above it. The mountain was formed by magma that emerged at sea level and mixed with the water, heating it to steam. The steam expanded in a great explosion, blowing the magma into tiny pieces that fell back into a newly created cone-shaped crater. In time the pieces pressed together. The now flattened, tilted crater was once much deeper, but erosion has worn down the rim. In time, ocean waves may wear Diamond Head back to the shoreline. Meanwhile, most of Diamond Head is preserved as a natural park. A neighboring island, the horseshoe-shaped Molokai, has also been cut back to sea level, but by landslide more than by erosion.

Another Honolulu volcano much like Diamond Head is the Punchbowl. Its crater filled with soil and

(Top) Middle and North Sisters are two volcanic peaks in central Oregon.
(Bottom) Diamond Head used to be the landmark for sailing
ships approaching Pearl Harbor from Asia or the West Coast. Today
Diamond Head's shape rises dramatically from the shore.

rock, creating a flat floor. The Punchbowl became a national cemetery after World War II, in the 1940s. The ground-hugging gravestones maintain the crater's natural beauty. This helps relatives and other visitors preserve a feeling of serenity as they honor the dead service men and women buried there.

A third extinct volcanic vent, Mount Tantalus, is different. Tantalus was formed on higher ground by an explosive eruption. Its crater is a cinder cone. At one time, lava flowed down the mountain to form a wide floor in the valley below. Today the valley is the location of the University of Hawaii at Manoa. The tree-covered Tantalus mountain itself is a natural park.

LAVA FLOWS AND LAVA FIELDS

A lava field is a place where lava once flowed, then cooled and came to a stop. The lava field formed below Mount Tantalus has been covered over by civilization, but many others are still out in the open and can be explored on foot.

All lava contains the same major ingredients, *silica* (or silicon dioxide) and aluminum oxides, but in different proportions. Also, lava takes many shapes as it flows from a volcano and cools from contact with the air. If you know how to "read" an old lava field, you can learn its history—the composition of the lava and the conditions when it was produced. Color, structure, and surface texture are all clues.

If the lava is light-colored, it is andesite, named for South America's Andes Mountains. It is the most common type of lava and has between 53 and 63 percent silica. If the lava is dark-colored, it is basalt and contains less than 53 percent silica. Kilauea and most Pacific and other ocean volcanoes produce basalt. Lavas containing the largest amounts of silica are called rhyolite, which is similar to granite, and dacite.

Silica behaves like an acid, so lava is also classified

Three types of lava (clockwise from upper left): rhyolite, dacite, and basalt

by its acidity—either acid, intermediate, or basic (alkaline). The more acidic the molten lava, the stiffer it is and the less likely to give off gases. The more gas it contains, the more violent its eruptions. Some lava fields contain both light-colored andesite and dark basalt lavas. You can be fairly sure that the basalt was produced by gentler eruptions than the andesite. Acidity also influences lava's flow rate and cooling pattern.

The lava's hardened structure can tell you what conditions were like as it cooled. If it is glassy (structureless), it cooled quickly. Lava that has a crystal structure and is coarse-, medium-, or fine-grained rock cooled more slowly. Andesite and basalt are usually fine-grained. Under unusual conditions, lava cools into forms that look like glassy beads, or diamonds, or even strands of hair.

Hardened lava's surface texture is the last major clue to its history. As it cools, most lava becomes either *pahoehoe* (pa HO ay HO ay) or *aa* (ah ah). Both names are Hawaiian, but the types can be found around the world. If the lava's surface texture is smooth and looks like strands of rope or is puffy like big pillows or charred marshmallows, it is pahoehoe. Large sections of the flow surface cooled, while molten lava continued to flow beneath it. This is similar to smooth ice that forms on top of a slowly flowing river. Kilauea is producing pahoehoe lava, and older pahoehoe fields can be found there too.

If the lava surface is rough, sharp, or jagged, it is aa. This texture tells you that some of the surface lava cooled as it flowed, rather than in large sections. Depending on how fast it flows and cools, aa lava may be big hunks or chunky balls. A field of prehistoric aa lava on Mauna Loa is one of perhaps too many examples, as volcano scientists who must climb through them can testify.

Lava Beds National Monument, in northern Cali-

(Top) Pahoehoe is lava that coils in large sections on the
flow surface and may look like large strands of rope.
(Bottom) Aa lava forms as a flow moves, so that the top
doesn't have a chance to cool in large sections.
Its surface is rough, sharp, or jagged.

fornia, contains ancient flows of both types of lava. It is located a mile high on the side of an active shield volcano, in the Medicine Lake highlands. Most of the flow is pahoehoe, but several aa flows, including one called the Devil's Homestead, can also be found. In addition, the monument has numerous cinder cones and hundreds of lava tubes. Many of the volcanic formations were used as forts and defenses by the Modoc Indians in time of war.

LAVA TUBES

Would you like to walk through a lava flow? It might be fun, especially if the lava isn't hot anymore. A lava tube or lava cave (a tube with a cavelike closed end) lets you take this walk. Lava tubes are cooled, hardened shells of lava that molten lava or hot gases continued to flow through. In a pahoehoe flow, the top hardens while the lava continues to flow underneath. If conditions are right, an entire tube of cool lava forms, looking very much like a big piece of sewer pipe on the inside.

Figure 9 shows how a lava tube is formed. Suppose the lava is flowing through a valley. The top of the flow is broader than the bottom, because the valley is V- or U-shaped. The center of the flow is the deepest and the least likely to cool, so it continues to move. The top, which is exposed to the air, and the shallower sides slowly begin to cool and become solid. As the central portion flows, it takes a very efficient shape, becoming rounded, like a cylinder. Finally, the flow slows down. Hot volcanic gases may fill in the gap between the lava stream and the tube's ceiling. But the tube remains. As the flow finally stops, the lava left in the bottom of the tube becomes a flat floor. In some lava tubes, the hot gases remelt part of the ceiling lava. As it drips, it may cool into long strips and hang from the ceiling like stalactites. Other formations are also possible, for ex-

(Top) Lava Beds National Monument in northern California
was the site of a war between the U.S. government and the
Modoc. The ranger stands in the stronghold of
Captain Jack, the Modoc chief.
(Bottom) Lava tubes are cooled, hardened shells of lava that
molten lava or hot gases continued to flow through. This one is
at Lava Beds National Monument, where many
tubes are open to the public for exploration.

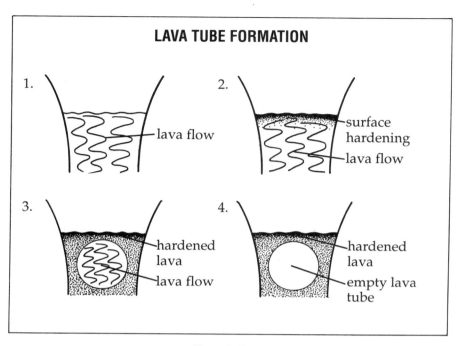

**Figure 9. How a
lava tube is formed**

ample, the "lavasicles" at the new Newberry National Volcanic Monument in central Oregon.

Lava tubes may be short or many miles long. One place to walk through a mile-long lava tube is at the Lava River Cave at the Newberry National Volcanic Monument. Another place to walk through lava tubes is at Lava Beds National Monument.

Tubes aren't the only unusual forms produced by eruptions. Tree casts and "fence posts" are two others.

LAVA CASTS

As lava flows through a forest, it may burn the trees to the ground. Sometimes the lava is very fluid and the trees are strong. Then the lava may surround a tree and cool, cooking the wood within it. What is left is a mold of the tree. The lava mold may even become a vertical lava tube, created by hot lava spouting up through the top of a cooled lava casing. The Lava Cast Forest in Oregon, near the Lava Caves, contains what are considered to be the most beautiful examples of lava molds in the United States.

Another place to see tree molds is the Lava Trees State Park, on Hawaii Island. Some are standing upright; others have been knocked over and are lying on the ground. A crater on nearby Kilauea even has some upside-down casts. A lava eruption in 1868 knocked over some huge treelike ferns, then buried them. Once the eruption was over, the molds were so perfect that scientists could peer down into them and tell what species of fern had been buried. Near another of Kilauea's craters, in an area called the Tree Molds, the lava continued to flow around the top of the molds, then hardened. These molds now look like wells, not towers.

"FENCE POSTS"

Sometimes new lava cracks as it cools. The result is lava posts, smooth shafts that look like huge wooden posts.

Imagine you were a nineteenth-century settler in east-central California. You have just discovered a large group of 60-foot-high posts made of rock. What would you call this formation? The real-life settlers decided to name it the Devils Postpile. Now a National Monument near the town of Bishop, this formation of three- to seven-sided lava posts was actually left over from an eruption 900,000 years ago. At the base of the upright posts is a pile of broken ones.

Long after the eruption, a glacier exposed the posts. Glacial action also leveled off the tops of the posts and polished them. The formation is the best-known of its kind in the United States. Similar formations exist in the British Isles and elsewhere.

The Lava Cast Forest, in the Newberry National
Volcanic Monument, contains the best collection of
lava molds in the United States. Here the photographer's
friend is standing in one of the molds.

The Devils Postpile National Monument, near Bishop, California

A smaller grouping of lava posts can be seen by people rafting down the Colorado River, through Arizona's Grand Canyon. Lava Falls Rapids in the canyon's west end is one of the wildest stretches on the river. It was created by eruptions about one million years ago. Lava blocked the river several times. Each time, the force of the water burst through. The lava posts can be seen above the rapids on the North Rim, along with a cinder cone.

WALKING ON THE MOON

Have you seen pictures of astronauts walking on the moon? The moonscape looks bleak and empty—just rocks and dust. No plants or other people or animals. Several volcanic areas on Earth have reminded people of the moon. There are plants and animals, but the landscape is dominated by strange cones, dust, and rocks. People rarely live in these places, and even traveling through them is hard work.

One of these areas is the crater of Haleakala volcano, on Hawaii's Maui Island. Haleakala (the house of the Sun) is active but hasn't erupted since 1790, so the crater's cinder cones (one is named Puu o Pele, or Pele's hill), huge red rocks, and other formations look ancient. You can drive to the rim, then hike down into the crater. Since the rim is almost 10,000 feet above sea level, even going downhill 3,000 feet to the crater floor can be challenging. Fortunately, hiking back out is a little easier. The exit is less than 2,000 feet above the floor.

Another mostly empty and high-altitude volcanic landscape is actually named for the Moon—the Craters of the Moon National Monument in southern Idaho. This area is volcanically active, but the last eruption was probably 2,300 years ago. The most recent cinder cone, named Sunset Cone, is over 11,000 years old.

The area's rugged formations of both pahoehoe

and aa lava have long resisted human settlement. From time to time, Native Americans have used its caves and tunnels for shelter or refuge. Nineteenth-century settlers could only go around it, and even then with great difficulty. Today, the trails to the monument's lava tubes and cones still challenge adventurers. However, you can drive to several viewing spots. The photograph on page 56 shows two interesting volcanic rocks found at the monument.

BANDERA AND EL MALPAIS

Some rugged volcanic areas go by another name: Badlands or, in Spanish, El Malpais. The names are used throughout the west, but one location may deserve it the most—New Mexico's Bandera Volcanic Fields. Its volcanic record stretches from 3 million years ago right up to the present.

Three million years of lava and rock is a lot of volcanic activity! The volcanic record of the Bandera Volcanic Fields, near the town of Grants in west-central New Mexico, was ancient when the first Native Americans arrived on the continent, perhaps as long as 40,000 years ago. The people who settled in this area undoubtedly witnessed the most recent eruption, because local legends describe "fire rocks." It took place 1,100 years ago in an area called McCarty's.

The heart of the field is a valley twice the area of Washington, D.C., preserved in 1987 as El Malpais National Monument. Cinder cones, lava tubes, and lava fields represent different stages of volcanic activity. They are interwoven with old Native American trade routes, pine forests, wilderness, sandstone arches dating from the days of the dinosaurs, and even a bat cave.

The first people to blaze a trail across the *malpais* were the ancient Anasazi. The trail then became a path between the Zuni and Acoma Pueblos. Today the same

(Top) Craters of the Moon National Monument in southern Idaho.
The last eruption was over two thousand years ago.
Two unusual lava rocks from Craters of the
Moon National Monument, in Idaho: Bottom left: a ribbon "bomb"
attached to a cinder; bottom right: a breadcrust bomb.

route is a hiking trail whose 6½ rugged miles cross four major lava flows (take your own water and keep a sharp eye out for the trail markers).

A chain of thirty cinder cones is probably the result of a magma hot spot. The largest cone is Bandera Crater, in another part of the site.

The monument's 17-mile-long lava tube and caves are best approached in an all-terrain vehicle. Hiking in over the 10 rocky miles from the paved road is more than most people want to attempt.

Why do all this hard work? People who go there want to experience both the beauty and the hardship of true wilderness first hand. They want to feel they are a part of it all—the forests and the wide-angle views, the volcanic remains and the clean air, the ancient trails and the Indian way of life.

CHAPTER 5

GEYSERS AND HOT SPRINGS

THE VOLCANO CONNECTION

Everyone knows what hot water is. It's ordinary hot tap water or cold tap water that's been heated on a range or in the microwave. There's also another kind of hot water—the kind that's cooked underground by the Earth itself. It comes up to the surface in three ways: geysers, which are spouts or fountains shooting up with explosive force; wells; and springs, which are surface wells.

Naturally heated water like this is called thermal, which means "hot." The word is used to describe water whose temperature is noticeably higher than the mean air temperature.

How hot is hot? It depends where the water is. In Europe, to be called thermal the water must be hotter than 64° F. U.S. waters are usually 15° F above mean air temperature. In colder areas, a thermal spring is often any spring that doesn't freeze. In hot climates, only a few degrees above mean air temperature is enough for classification as thermal.

Thermal waters are rather uncommon, so many people think they must have special qualities. In earlier times, people thought they had magic or healing powers. For thousands of years, and in many countries, these waters have been used to treat illness. In modern times, some of them are being used as natural sources of commercial energy.

Thermal water gets its heat from the Earth's magma, so it's not surprising that many volcanic areas have thermal waters. Volcanic activity isn't required, though. Some geologic formations allow water to seep far into the ground to be heated by coming near magma that isn't moving.

ORIGINS OF GEYSERS AND HOT SPRINGS

Water found underground in permeable layers of rock (for example, sand and gravel) is called an aquifer. Some aquifers are near the surface, and some are deep within the Earth. If this water is hot, it has come in contact with magma or with rock formations heated by magma. A geyser occurs when magma warms underground water enough to make it expand and burst upward periodically. Sometimes the heated water remains underground in deep wells. Sometimes it may bubble quietly to the surface as a hot spring.

Thermal water is also found in nonvolcanic rock formations where faults exist. The famous resort at Hot Springs, Arkansas, uses water that makes a long round trip through surface faults to the Earth's heated interior and back. In some cases, the round trip takes many years, or even thousands of years.

Some thermal sources are mysteries. Scientists haven't been able to trace them to either faults or volcanic activity. The southern California desert spring at Twentynine Palms is an example.

Some rock formations may contain only water. Others hold both water and steam, or steam alone. Thermal wells can also be engineered in hot but dry rock. Drilling holes in them allows cool water to be pumped in, heated, and pumped out for commercial uses like generating electricity.

Over half of the fifty states have at least one thermal water source. There are more than 1,100 in all. Figure 10 shows where they are. Most are in volcanic regions—California, Idaho, Nevada, and Wyoming—but not all volcanic regions have large numbers of thermal sources. Hawaii has only ten, two on the island of Maui and eight on Hawaii Island. And Alaska, with its many volcanoes, has only seventy-nine.

Many active volcanic areas in Washington, Oregon, California, Nevada, New Mexico, Alaska, and Hawaii have thermal waters. The most spectacular ones are found around Yellowstone National Park, which is in northwestern Wyoming and nearby Idaho and Montana.

VOLCANIC THERMAL SOURCES

Most Americans know about one feature of Yellowstone National Park, even if they've never been there—Old Faithful Geyser. Several times every day it faithfully spouts up fountains of water, delighting the visitors who have gathered around waiting for the eruption. Yellowstone Park is also famous for its bears and other animals and for its mountains, lakes, and rivers. That's only what's on the surface.

Yellowstone is just as exciting underground. The National Park is on a huge, two-part, volcanically active geological formation—Yellowstone Plateau and the Eastern Snake River Plain. Yellowstone Plateau, the heart of the park, is actually a huge caldera, a flat-bottomed, gently sloping crater almost 30 miles wide and 50 miles long. Not far below the surface is a huge pool of magma that has been active for the past 2 million years. Scientists aren't sure yet why volcanic activity occurs here, but are studying four possibilities: a thin area of crust moving over a hot spot of magma or radioactive material; a major break in the lithosphere; excessive heat created by friction at the bottom of the lithosphere (like the heat you make when you rub your

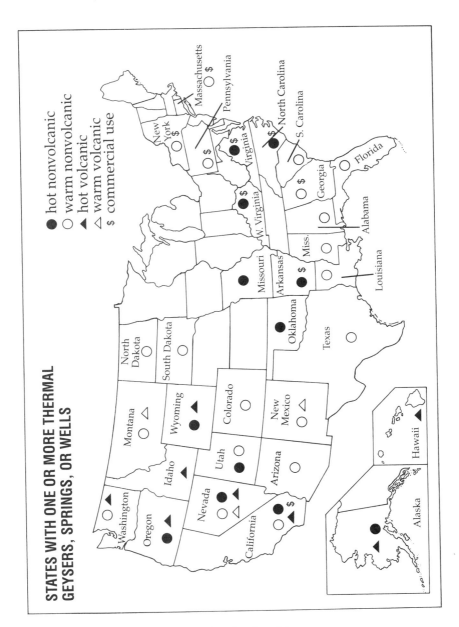

Figure 10. Map showing states
with one or more thermal
geysers, springs, or wells

Old Faithful in Yellowstone National Park

hands together); and magma moving upward under the entire north-central United States.

The volcanic activity began about 14 million years ago in the second half of the formation, the Eastern Snake River Plain, southwest of Yellowstone Park. The activity has been creeping northward at about an inch a year, perhaps because the North American Plate is moving west. The path is marked by the remains of gigantic eruptions. The last of these, about 630,000 years ago, created the Yellowstone caldera. The smaller Island Park caldera is southwest of it.

The magma chamber is over 12 miles thick. Its roof is, at most, only 3 miles below the surface, and in some places the magma is only a mile or so beneath the surface. This means the rock above the magma is very hot—1,300° F to 1,650° F. Above this rock are large bodies of water.

The water heats and expands, pressure builds up, and the water explodes through fractures to the surface, producing geysers. Old Faithful isn't alone. It has lots of companions: Jupiter Terrace, Norris Geyser Basin, Mammoth Hot Springs, Shoshone Basin, and Heart Lake Basin.

NONVOLCANIC THERMAL SOURCES

The Appalachian Mountains, stretching from Maine to Georgia, are ancient, between 170 million and 3½ billion years old. In their long lives, they have become heavily faulted. The region receives lots of rain and snow. Some of it seeps down through the faults until it comes close to the magma, heats up, and is stored within the rock. Warm Springs, Georgia, and Hot Springs, Virginia, are two well-known Appalachian resorts built over thermal water supplies.

The Ozark, Arbuckle, and Ouachita mountains of Missouri, Arkansas, and Oklahoma are younger—only 270 to 600 million years old—but they are largely limestone, which is ideal for collecting water underground. Where the rock is faulted, the water moves down far enough to become hot, as at Hot Springs, Arkansas.

The sandstone formations of North and South Dakota are even younger, at 135 million years. They have many deep aquifers of warm (not hot) water, 20° F to 25° F above the temperature of surface water.

Which are the thermal water champions? Warm Springs Creek, in the Rocky Mountains of Montana, is the largest natural stream of thermal (68° F) water in the United States, running at 80,000 gallons a minute. The biggest hot spring is thought to be at Big Horn (also called Thermopolis, meaning thermal city), Wyoming. Its water is 135° F and gushes out at 2,600 gallons a minute.

Even heavily volcanic states like Alaska, California, and Nevada have some nonvolcanic thermal waters. On Alaska's Seward Peninsula, hot springs originate in faults in ancient rock, not in younger, volcanic rock.

DO HOT SPRINGS HAVE ANY HEALTH BENEFITS?

If you go into the cosmetics section of a department store, along with lipstick, hair gel, and cologne you will see jars of mud. This isn't just backyard mud. It is expensive mud, said to have special qualities, from one of several hot springs that have been famous since ancient times.

For thousands of years, people have believed that hot springs and their products, like mud, can cure illness. They have thought the waters can restore energy and make people younger.

A hot spring in Turkey was described by Homer in the *Iliad* almost 3,000 years ago. Southern Italian hot springs famous in Roman times, 2,000 years ago, are still popular today.

The city of Bath, England, was literally named for

its hot spring, which also has been popular since Roman days, and perhaps earlier. The name of Baden Baden, a hot spring resort in Germany, means baths baths. In Japan, ancient hot springs, complete with bubbling hot mud, are still popular; many people, including Olympic athletes, use them as energy restorers. Resorts like these are often called spas, from the name of a Belgian hot spring, Spa.

Before modern medical treatment, people with many different ailments crowded into the hot baths for hours at a time. They often spread diseases, even as they thought they were being cured. Drinking the waters was another part of the "cure." Even today, many Europeans believe in spending a week or two at a spa. They drink the waters, thinking they will improve the condition of their livers and their health in general.

The United States has also had its share of "medicinal" hot springs. Many were known to Native Americans before Europeans settled here. Some springs in Virginia and what became West Virginia were turned into spas by eighteenth-century colonists and have been popular with Americans ever since.

The resort at Warm Springs, Georgia, became world famous in the 1930s as a rehabilitation center for people who had poliomyelitis, a dreaded paralytic disease. (Vaccinations have prevented epidemics since the 1960s.) The only known polio treatment at the time was heat, used to relax the muscles and make physical therapy possible. Swimming in the warm waters was one way of applying heat. President Franklin D. Roosevelt, who had polio, went there regularly in the 1930s and 1940s.

Perhaps the best-known health spa in the United States is in Hot Springs, Arkansas, established in 1832. Each year, several million people visit Hot Springs National Park for vacations, and thousands use its forty-seven springs for health reasons.

Do the waters of thermal springs really bring good health? It depends on how you define "good health."

There's no doubt that soaking in a hot tub is very relaxing. It relieves stress, even if the water only comes out of your home hot water heater. On vacation, especially in a beautiful setting, it can seem even more helpful. Some people even like to soak in warm mud.

Are hot springs Fountains of Youth? No. But the sheer sense of relaxation they provide often makes people think of them as such. Often visiting a spa means "taking the waters" and following a special diet. And eating healthful food can make people healthier, if not younger. Long-term health benefits will result only if the people continue to reduce stress and eat healthful food after they return home. Even the cosmetic benefits of the baths and mud last for only a few hours. There is no evidence of any long-term benefits.

How about drinking the waters? Isn't that healthful? People sometimes drink thermal waters for their mineral content, but ordinary water also may contain minerals. The hot water at Hot Springs, Arkansas, for instance, is the same as water from cold springs nearby. Both the hot and cold waters contain small amounts of minerals, but also slightly higher than average amounts of radon, a radioactive element. Both the minerals and the radon are considered safe to drink. The hot water tastes different from the cold, but only because heat affects the flavor of some minerals.

The content of thermal water is different at each source. Some thermal water contains useful minerals like calcium and magnesium. Some contains carbon dioxide, which makes it bubbly. That may feel or taste good, but it doesn't have any other advantage. Hot springs in volcanic areas may contain hydrogen sulfide, the same gas produced by solfataras. Water containing hydrogen sulfide smells like rotten eggs, but many people think it's good for them—for the same reason some people think that bad-tasting medicine must be good for them.

Anyone who plans to drink thermal water should be careful to avoid sources that contain substances in concentrations high enough to be toxic. Examples are the elements arsenic, boron, and radium.

In the early twentieth century, "radium cures" were promised from some thermal waters that were slightly radioactive. Cancer and a variety of physical ailments were said to improve after bathing in the water or drinking it. People even added the word *radium* to the names of some springs to cash in on the craze. Radium Hot Springs, New Mexico, is an example. The promised cures were fakes. There's no evidence that small amounts of radium in drinking or bathing water can cure any ailment.

COMMERCIAL USES OF THERMAL WATER

Health resorts were the first commercial use of thermal waters. Today, some thermal sources are being used for fish and shellfish farming. In at least one place, Paso Robles, California, the water is used twice—once to raise catfish and a second time for irrigating farmland. Some water provides heating for homes, barns, and greenhouses. In a few cases, it is being used in manufacturing. The biggest use is for generating electricity.

Some countries make great use of their thermal waters. Italy began generating electricity from volcanic steam in 1904. Since the 1930s, Iceland has heated homes and even outdoor swimming pools with it. Japan, the Soviet Union, Hungary, and New Zealand are other countries that put their hot water to work for them.

In the United States, there has been only limited use of thermal energy as a fuel. It has seemed too expensive, or other fuels have been cheaper or easier to get. Many thermal sources are in untouched wilderness areas, where all development is forbidden. Elsewhere, zoning laws prevent development. Some water can't be used because it contains salt, boron, or other chemicals that could become environmental pollutants if removed.

Environmental and economic factors may make thermal energy more attractive in the future. The burning of fossil fuels (oil, coal, and gas) adds carbon dioxide to the atmosphere. This may be speeding up global warming—the greenhouse effect. It also creates air pollution, including acid rain. Nuclear power brings the hazards of long-term storage of radioactive wastes.

Thermal power may be a clean fuel whose time has come, especially in the western states. California already gets 7 percent of its electricity from thermal sources.

GEOTHERMAL HEAT AS A SOURCE OF ENERGY

Underground heat may be stored in water, steam, or rock, so engineers use the general term *geothermal*—meaning earth plus heat—to include all three.

California is the center for geothermal energy in the United States. The Geysers, near the town of Calistoga, about 60 miles north of San Francisco, has been used for many years to generate electricity. Four other California locations are under serious consideration, as Figure 11 shows. All result from volcanic activity.

Heated water already runs the turbines that generate most electricity. (Hydroelectric power relies on the force of water to run the turbines.) Usually heat energy from the burning of coal or oil, or from nuclear power, raises the temperature of cold water high enough to drive the turbines. In theory, geothermal formations are just another source of heat energy. In reality, ordinary turbines must be modified before they can work efficiently with geothermal heat. This is be-

A geothermal plant at the Geysers, near the town of
Calistoga, about 60 miles north of San Francisco

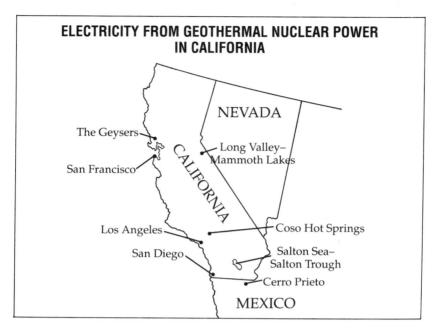

ELECTRICITY FROM GEOTHERMAL NUCLEAR POWER IN CALIFORNIA

NEVADA

The Geysers

Long Valley–Mammoth Lakes

San Francisco

CALIFORNIA

Los Angeles

Coso Hot Springs

San Diego

Salton Sea–Salton Trough

Cerro Prieto

MEXICO

Figure 11. Electricity is obtained from geothermal power at several locations in California.

cause geothermal sources produce less heat than burning fuels. But three North American success stories show that geothermal energy can be practical.

The Geysers was developed in the nineteenth century as a health resort. It doesn't have any real geysers, but it does have several fumaroles, some of them named Smokestack, Steamboat, and Safety Valve. In the 1920s, developers began drilling holes, hoping to find pressurized steam for generating electricity. They succeeded, but no one wanted to buy their power.

In the 1950s, others tried again, this time successfully. Today, steam from The Geysers field generates 900 million watts, or 900 megawatts (*mega* means million), of power, enough electricity for over 900,000 people. It is one of the few pressurized steam sources in the world. The power is sold to the public utility serving San Francisco and other parts of northern California. The field's production could be expanded to 2,500 megawatts. One megawatt is enough energy to light 10,000 100-watt light bulbs.

A smaller field, farther south in California, is Coso Hot Springs. Its volcanically heated water is used to generate electricity for the China Lake Naval Weapons Center.

The third success story isn't actually in the United States but just 30 miles south of the border in Mexico. A geothermal field called Cerro Prieto is the biggest developed field of hot water in the Western Hemisphere, producing over 150 megawatts of electricity. It is part of a volcanic formation called the Salton Sea volcanic field that extends into the United States. This is a very young volcanic area, with two volcanoes located in the Salton Sea itself. An eruption may have taken place any time within the past 16,000 years.

Existing hot water wells in the Salton Sea area are being studied for cooling, rather than heating, because of the climate. In fact, the entire area is sometimes called the Persian Gulf of geothermal energy. This is because it is believed to have as much energy as all the oil deposits in the real Persian Gulf, in the Middle East. Much of the water is extremely salty, allowing it to hold much more heat than most thermal water does. In some places the water temperature is 500°F.

Volcanism has fascinated people since ancient times. Even today's science hasn't solved all the mysteries of volcanic activity. How do volcano scientists go about their work? This is the subject of the next chapter.

STUDYING
VOLCANOES

Eighteen-year-old Pliny and his mother ran down the street of the seaside town, choking on the gas-filled air. Behind them, fiery lava poured from the volcano. Panicking people called out to each other, or prayed for divine help. Others in the street stumbled and fell, unable to continue. Pliny and his mother took refuge in a field, to avoid being trampled on the road. They survived, but Pliny's uncle wasn't so lucky. He reached the beach, but the water was too choppy for boats. Choking on the sulfurous air, he clutched at his chest and fell, victim of a fatal heart attack.

**—adapted from the letters of
Pliny the Younger to Tacitus, A.D. 79**

The oldest written histories of volcanoes come from ancient Rome. The first observations were made for religious reasons. Perhaps praying to the volcano or its god would keep them safe, many Romans believed. Pliny's story is based on his description of a major eruption of Mount Vesuvius. In a few hours, lava and toxic gases destroyed two fashionable resort cities—Herculaneum and Pompeii—burying many people alive and suffocating others. A teenaged scholar, Pliny wrote the first known objective observation of an eruption.

Over the centuries, observation has changed into scientific investigation. Today's *volcanology*, the science of volcanic activity, involves field study, computer analysis of data, and the major goal of being able to predict eruptions far in advance.

THE BEGINNINGS OF VOLCANOLOGY

Because of Italy's many volcanoes and a long written history of them, it's not surprising that the first scientific volcano studies were done in that country. The first modern volcano observatory was established near Mount Vesuvius in 1841, and has long been a world-renowned research center. Many countries have volcano observatories or study centers, including Japan, the Soviet Union, the South Pacific nation of Papua New Guinea, the Philippines, and New Zealand.

Two American scientists, Dr. Thomas A. Jagger and Dr. Frank A. Perret, established volcanology in the United States during the early 1900s. Dr. Jagger set up this country's first observatory, on Hawaii Island in 1912. A second U.S. observatory, in Vancouver, Washington, was established in 1980, especially to study Mount St. Helens. Now its scientists study many volcanic areas. Anchorage, Alaska, is the home of a third U.S. observatory. Also, the Smithsonian Institution has established a reporting network, called the Global Volcanism Network (until 1990 known as SEAN, the Scientific Event Alert Network). Scientists, airplane pilots, and others report any volcanic activity they see or detect. The information is available to scientists and other interested people around the world.

Scientists study volcanic areas in several ways. They look at evidence from past eruptions so they can construct a life history of the volcano. They measure magma behavior and its effects on the Earth's surface. And they observe eruptions while they occur.

DISCOVERING A VOLCANO'S LIFE HISTORY

Each volcano or volcanic area is unique, and each one reveals its story in a different way. It might be through written descriptions, traditional legends, or the belongings of people who once lived there. Or it might be through the chemical composition of rocks and other materials, or the remains of trees and other plants that grew there.

Life history often means the written historical record, but "historical" means something different for each volcano. Written records of Italian volcanoes go back almost 3,500 years. The written record of the Cascades volcanoes is less than 200 years old.

Sometimes the stories and legends, like those about Madame Pele or the gods of Crater Lake, provide valuable information. So do the belongings of local people. During an eruption, the lava or ash may damage personal or public property, as shown in the photograph on page 75. The location, style, and stage of technology of these belongings can help scientists estimate the date of the eruption that buried them. This method helped scientists decide that an eruption at Bandera Lava Field, in New Mexico, took place around the year 900.

When there is no direct way to tell when an eruption occurred, scientists can discover the dates within 100 years by measuring the amount of carbon-14 present in an object from the site that contains carbon. The less radioactive carbon-14 in relation to stable carbon-12, the older the object is. The carbon-14 method, which is accurate for the past 10,000 years, has been used to date many eruptions in the Cascades.

Natural remains, such as tree rings and other re-

During an eruption, flowing lava may destroy personal
or public property. This lava is from Kilauea.

mains of plants and animals, are good clues to the dates of eruptions. Dendrochronology (literally tree dating) is a reading of the growth rings in a tree trunk. Each ring marks a year in the tree's life. The distance between rings can tell scientists whether growth was fast or slow from year to year. Volcanic activity, for example, affects tree growth. If scientists already have an approximate date for an eruption, the tree ring history can help them be more accurate. This method showed that Mount St. Helens erupted in about 1800 and that Sunset Crater, in Arizona, erupted in 1064.

In parts of the far north, a plant called lichen grows on rock at a regular rate. Disruptions of this growth rate can be measured by a method called lichenometry. If the lichen is growing on lava, the growth disruptions can tell scientists the date the lava made its appearance.

The amount of pollen trapped in lava or ash may tell which season of the year the eruption occurred. The sediment on lake floors also changes with the seasons, and its layers can be read like tree rings. The remains of such layers, called varves, were used to date an eruption of Mauna Kea, Hawaii, from about 1,800 years ago.

Changes in the Earth's magnetic field, changes in the thickness of rock surfaces, and other geologic information also help scientists learn when eruptions occurred. Uranium fission tracks are another source of information. Some rock formations contain the element uranium. When an atom of uranium fissions (splits in half), it leaves a track in the rock. The larger the number of fission tracks, counted under a microscope, the older the rock.

Layers of ash can also be measured to provide eruption dates. This method was used to date an eruption at Mount Baker, in Washington, from 7,500 years ago.

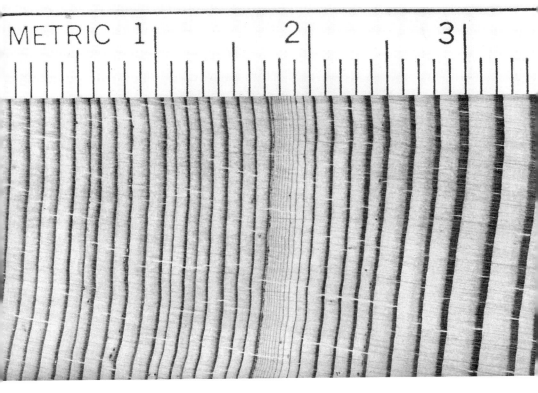

A volcanic eruption can defoliate nearby trees, slowing or stopping growth temporarily and leaving a record in the pattern of rings. The photo is of rings in a 600- to 650-year-old Douglas fir affected by an eruption of Mount St. Helens in the late fifteenth century. The dark, wide rings (left) created by normal growth abruptly narrow at eruption, then widen again as the tree recovers. The rings on the right indicate even stronger growth after the eruption, as a result of less competition from other trees since many were destroyed.

MEASURING MAGMA ACTIVITY

One very good sign of magma activity is a bulge, or uplift, in the surface of the Earth. The Goat Rocks bulge was an important pre-eruption sign at Mount St. Helens in 1980. Volcano scientists use electronic instruments called tiltmeters to measure these changes. Some of them are so sensitive that they can detect an uplift of 1/25th inch along a 60-mile line. In water-filled calderas, changes in water level can also indicate active magma.

Volcanic earthquakes are studied the same way other earthquakes are, with seismometers. These instruments measure the quake-caused pressure waves that travel through the Earth. Scientists place a network of seismometers around a volcanic area. They then combine the information from all of them to create a three-dimensional picture of what's happening down below. Mount St. Helens, Yellowstone, Kilauea, and several other U.S. volcanic areas have seismometer networks. If the volcanic area is at a plate boundary, the instruments can also tell how fast the plates are moving.

Are volcanic and nonvolcanic earthquakes related? Sometimes they are. Magma moving underground can also cause ordinary earthquakes. Ordinary earthquakes can sometimes trigger an eruption. Volcanic and ordinary earthquakes occur in the same general locations, though earthquake areas are bigger. In addition, volcanoes about to erupt create a special type of earthquake, one with a distinct energy pattern called *harmonic tremor.* This was the vibration pattern detected at Mount St. Helens that showed an eruption was due at any time.

Ordinary earthquake waves are usually created as the crust rips apart along a fracture or plate boundary such as the San Andreas fault. Harmonic tremor is different. As in music, "harmonic" refers to the relationship among the waves. There are several theories to explain harmonic tremor. It may be a vibration pattern set up in the magma channel to the surface. This is similar to the vibration pattern created when a musician plays a wind instrument. Or it may occur when moving magma forces the rock to fracture. The photograph on page 80 shows the difference between ordinary earthquake and harmonic tremor patterns, as recorded on a seismometer.

The timing and location of ordinary earthquakes and volcanic activity in any area are usually very different. Studies around the world show that an area's largest earthquakes and its largest eruptions are almost always in different places and that they occur at different times, sometimes decades apart. For instance, a powerful earthquake struck Anchorage, Alaska, in 1964, the worst there in history. But the greatest volcanic activity in the region took place half a century earlier, in 1911–12.

A volcanic eruption is as busy a time for scientists as it is for the volcano. The tiltmeter and seismometer readings have already alerted them. When harmonic tremor tells them that eruption time is near, they swing into action. Some take off in airplanes and helicopters to record the volcano with photographs and video. They also make special infrared (heat wave) measurements and pictures, called thermographs. These show heat patterns on the surface caused by the hot magma moving underground.

If the eruption is taking place underwater, scientists use special listening devices called hydrophones ("hydro" means water). These measure shock waves that pass from the volcano through the water. Sometimes scientists use research submarines to see the results of such activity first hand. Other scientists use lasers to measure ground movements from safe distances. Meanwhile, some are analyzing the contents of the volcano gases with instruments called spectrome-

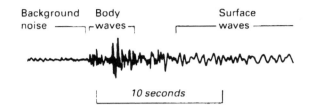

Background   Body   Surface
noise ——————┐┌ waves ┐   ┌———— waves ————

|——— 10 seconds ———|

Seismogram of typical earthquake recorded by seismometers placed near Mount St. Helens.

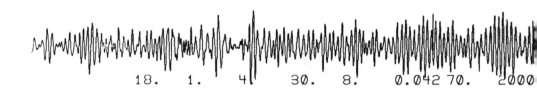

18.   1.   4   30.   8.   0.042 70.   2000

22.5 SFC. |

Seismogram of typical harmonic tremors recorded by seismometers placed near Mount St. Helens.

**Seismograms of an
ordinary earthquake and
of harmonic tremors**

ters. If the ash is very thick, scientists use radar to record surface changes through it. Later, they make these pictures more "readable" with computer techniques. Still more pictures and information come from weather satellites, as their orbits bring the area into view.

Once the eruption is over, or during quiet periods, volcanologists go to the eruption site and collect samples. They also measure changes the eruption has caused in surface features (such as those caused by landslides), river courses, or height, for example. The scientists study all their information for many weeks or months as they try to understand just what has happened.

HOW TO BECOME A VOLCANOLOGIST

Volcanology is part of a field known as the earth or planetary sciences. It's what is called "interdisciplinary." This means that people from several study disciplines are involved. Earth scientists such as geologists, oceanographers, meteorologists (weather scientists), and geophysicists may all take part in volcanic research. Planetary scientists perform similar studies, but their subjects are the Moon, other planets and their moons, and the asteroids and comets. This field can be important for volcanology, because each planet and moon seems to have matured differently. The presence or absence of volcanoes on them helps explain volcanoes and plate movements on Earth.

Becoming a volcano scientist involves obtaining a bachelor's degree and, for many, graduate work leading to a doctorate. About 500 U.S. colleges and universities offer bachelor's degrees in the earth and planetary sciences. The major may be in this general category or in a specific science such as geology. Dozens of public and private universities in all parts of the country have graduate programs in one or more of these disciplines. Some universities have specialties in volcanology.

Besides taking classes and studying, graduate students do field work at volcanic sites and interpret the information collected, almost always with the help of a computer. In the early part of graduate school, the student works under the direction of a professor or more advanced student. A doctorate requires original research. An advanced student is expected to take the lead, working under a professor's supervision.

Volcanologists may specialize in the volcanoes of one area, such as the Pacific Northwest or Hawaii. But the effort is international. Research in one part of the country or in one nation helps scientists all over the world with their work.

EMPLOYMENT

Where do volcanologists work? There are three main kinds of jobs for people with doctorates.

One is teaching in a university. Most universities encourage their faculty members to do a lot of research. In fact, this is often more important than teaching classes. Research usually involves working with students and other faculty at the school and on field trips. As with all scientific research, results must be written up as articles and published in professional journals, so others can try to duplicate them, to show if they are correct.

The second area of employment is with the U.S. Geological Survey (USGS), a part of the U.S. Department of the Interior. The USGS operates the three volcano observatories in Hawaii; Vancouver, Washington (a suburb of Portland, Oregon); and Anchorage, Alaska. The Vancouver observatory is named for David Johnston, the volcanologist who lost his life during the Mount St. Helens eruption. Such a death is very rare in volcanology, but some physical risk is connected with volcano research.

The third area of employment is with state governmental agencies in the volcanic states. Examples include the California Division of Mines and Geology and the Arizona Bureau of Mines.

One reason to study volcanic areas is purely scientific—to gain knowledge of the Earth and how it works. But there's a practical reason, too. Volcanism can affect the environment we all live in. The next chapter discusses the effects of volcanoes on people's lives, jobs, and health.

CHAPTER 7

VOLCANISM AND THE ENVIRONMENT

Mount Erebus, the Earth's southernmost active volcano, is the only one in Antarctica. Like many volcanoes, it gives off gases, but Mount Erebus also produces something unique—flecks of gold. There isn't enough gold to accumulate as money, but it is valuable in a different way—for pollution studies. The atmosphere's ozone layer is getting thinner and has a hole in it, centered over Antarctica. With less ozone, increased amounts of cancer-causing ultraviolet radiation will reach the entire Earth. The ozone thinning is partially caused by industrial chemicals, but scientists believe volcanic gases also play a role. The gold from Mount Erebus helps scientists track the volcano's contribution to the ozone destruction.

Volcanoes affect the environment and human life in many ways. Sometimes the volcanic effect has been helpful, as in fertilizing Hawaii's soil. Sometimes the effect is harmful, like ozone destruction. On a smaller scale, the 1989 eruption of Alaska's Redoubt forced the temporary grounding of air traffic, because the ash could damage the airplane engines. And sometimes the effect has shaped the very development of the Earth and of life on it.

Understanding volcanoes helps people learn how today's Earth was shaped over time. It also explains how they affect our health, our weather, and even the way we earn our living.

WHAT VOLCANOES TELL US ABOUT THE EARTH

Volcanic activity is an important link among the Earth's three great recycling systems—the plate tectonic cycle, the rock cycle, and the atmospheric cycle. Figure 12 shows how they fit together. Without plate tectonics, most volcanic activity would not occur. Erupting magma was the foundation of the continents, and more recent crust layers were also produced by volcanoes.

The Earth's surface, in turn, is the focus of the rock cycle. Magma activity brought metals from the Earth's core to the subsurface, where they concentrated. Changes in the crust later brought them to the surface. In this way, volcanism set the scene for the Gold Rush of 1848 and the modern metal mining industry. At other times, the heat and pressure of the underground magma changed the structure and composition of the surface rock.

Volcanic gases were important contributors to the formation of the young Earth's atmosphere, supplying the basic environmental gases. Ammonia, carbon dioxide, and water vapor in the gases contributed nitrogen, oxygen, and water to the atmosphere, which in turn permitted the development of living organisms. It was decaying organisms, plus atmospheric weathering and further volcanic activity that formed the top layer of the Earth, the soil.

Volcanic action plays an important role in another part of the atmospheric cycle—water. Volcanoes recycle water in the form of steam. Scientists now think they may be an important source for renewing the total

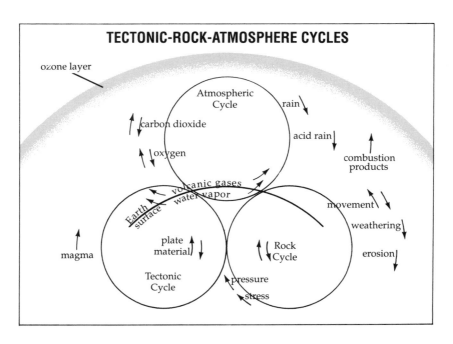

Figure 12. Tectonic, rock, and atmospheric cycles, showing how these three major environmental cycles interact

mineral content of sea water. The ocean floor contains many volcanic areas. As with surface volcanoes, some erupt explosively. In these cases, the volcanic gases bubble the minerals widely through the water.

Other volcanoes spout sea water that has filtered downward far enough to be heated by magma. As the water warms up, it expands, shooting upward. Loihi, Hawaii Island's undersea volcano, has such water spouts. On its round trip, the water picks up various minerals. It usually shoots up gently, because of the ocean pressure holding it down, so the minerals remain nearby on the ocean floor. The warmth and minerals provide a friendly environment for colonies of sea creatures. In some places, these are very tiny animals, but in parts of the Pacific, they are huge tube worms and giant clams.

A computer-generated "fishnet" image of Loihi,
Hawaii Island's undersea volcano

VOLCANOES AND CLIMATE

Even the largest eruptions affect only a small part of the Earth's surface, but they can have worldwide effects on climate. In 1815, a volcano named Tambora, in the Pacific nation of Indonesia, had an eruption so large that its clouds of ash, dust, and gases circled the Earth. The volcanic clouds were what scientists call a dust veil, a curtain screening out enough of the Sun's heat and light to change the Earth's weather for the next several years. Because of the year-round cold and snow that resulted, 1816 became known as The Year Without a Summer, or "Eighteen-hundred-and-froze-to-death."

How could the dust veil from one eruption, even a large one, change the weather of so much of the planet? The answer is sulfur. Most volcanoes contain large amounts of it and give it off as volcanic gases such as sulfur dioxide. Sulfur products from volcanoes can block out the Sun's heat and light. In fact, they behave the same way sulfur pollutants from cars and power plants do.

Tambora's clouds were rich in sulfur. Mount St. Helens's 1980 clouds didn't have much sulfur, so its eruptions affected the weather for only a short time. The years-long eruption of Kilauea has produced lots of sulfur. Because it is much smaller than either Tambora's or Mt. St. Helens's, its effect has been only local. In Hawaii's moist climate, however, the sulfur has turned to acid rain, a substance also created from industrial pollution. In the northeastern United States, acid rain is destroying vegetation and may be a health risk. The acid rain caused by Kilauea is as great as that caused by industry in other areas of the country.

VOLCANOES' EFFECTS ON PEOPLE

The Mount St. Helens eruption in May 1980 was one of the most watched volcanic events in history. For one thing, the eruption was very big. It took place in a modern nation, with experts and equipment nearby to study it. For another, Mount St. Helens was the first volcano to erupt in the lower forty-eight states in sixty years. Its location, relatively close to a large city, made the eruption a natural for TV coverage. But for the people of the region, the eruption was more than a media event. It affected their health and the local economy.

Kilauea's ongoing eruption has been watched continuously, mostly by scientists and people living nearby. Hawaiian eruptions are usually called quiet, because they're not explosive. Kilauea makes news periodically when its "quiet" lava still manages to burn through subdivisions, destroying homes and changing people's lives. And its quiet air pollution—people call it vog (*volcano fog*)—may be changing their health.

THE HEALTH EFFECTS OF VOLCANOES

The 1980 Mount St. Helens eruption killed approximately sixty people. For many thousands of others, the greatest harm came from the ash-filled air.

Ash is corrosive; it eats away at living tissue. The ash from Mount St. Helens created breathing problems for people with asthma and hay fever. The high-sulfur-content vog from Kilauea is also corrosive. Residents blame it for various breathing problems, such as asthma, sinus trouble, and allergies. Health experts are studying both the atmosphere and the people to learn if this is true.

Kilauea's lava may create another health problem as it enters the sea at Kalapana. The sulfur in the lava could combine with the water, creating acid steam. Civil defense officials are ready to evacuate people if the air becomes too toxic.

Cleanup after an eruption is a major job. Loggers, farm workers, and highway crews downwind of Mount St. Helens worked for weeks to clean up the tremendous amount of fallen ash. They all had to breathe air that was much more polluted than would be allowed in a factory or a mine. Some of the workers ended up with permanently damaged lungs. The ash and volcanic gases irritated and sometimes injured people's eyes, too. In addition, hospital emergency rooms reported a rise in car accident injuries. These were probably caused by poor visibility, bad road conditions, and stress.

An eruption, like any major emergency, is very stressful. At Mount St. Helens, some people nearby had to be evacuated quickly. Downwind of the volcano, farther to the east, the ashfall disrupted work, school, and other normal activity. Mudslides buried roads and property, as shown in the photograph on page 90. All contributed to people's temporary emotional problems. Family arguments, nightmares, and fear of another eruption were high on the list of problems. Some people became very passive, not wanting to get up in the morning to go to work or school. For most, the problems cleared up as the ash was removed and life became normal again; but for a few, the problems went on for a year or more.

In Hawaii, people have watched as their homes and the possessions of a lifetime were gobbled up by a slow-moving wall of lava. Others had time to move their belongings to safer areas. A church was even moved several blocks away, out of the lava's path. Some people were angry. Others were philosophical, saying "the land belongs to Madame Pele."

VOLCANOES AND PEOPLE'S FINANCES

One great cause of stress and emotional problems is loss of income. If a volcano hurts the local economy, people's jobs and income are harmed, too.

Eruptions can be both good and bad for the local economy. Of course, they destroy. During the first year

Cleanup after a volcanic eruption is a major job. Five feet of mud in this trailer park was a result of the Mount St. Helens eruption.

after the Mount St. Helens eruptions, the area's tourist industry lost money. People canceled reservations because they feared another eruption, or they decided to go elsewhere because so much natural scenery was destroyed.

Volcanoes can also be tourist attractions, bringing people and money. People came to witness Mount St. Helens in action. Afterward, they came to see what had happened. They still do. Such interest often lasts for decades or longer. Lassen Peak and its surrounding national park are popular with tourists seventy years after the eruption. Crater Lake is a major tourist attraction and probably will continue to be one.

In Hawaii, eruptions have drawn tourists for well over a century. The still-popular hotel called Volcano House, near Kilauea, opened in 1866. One of its first guests was the famous American writer Mark Twain. Today people fly in from other islands and jam the nearby roads. Charter planes fly sightseers right over the eruption zone. Some airlines even have lists of people who automatically are booked whenever the eruption becomes unusually active.

Volcanoes and tourism go together in many parts of the world. In Italy, scientists wanted to protect a town during the 1971 eruption of Mount Etna by changing the course of the lava flow. But local merchants objected. They feared sightseers and tourists would stop coming, which would harm the area's economy.

VOLCANOES, NATURAL RESOURCES, AND AGRICULTURE

The immediate effect of an eruption on agriculture is the loss of crops. Corrosive ash, gases, and heat make them wither. The sheer force or weight of the volcanic products may simply bury the crops. On the other hand, lava and volcanic mud contain minerals that fertilize the soil. In Hawaii, once the lava has cooled, it makes the soil fertile for a long period of time, fertile enough for vegetables to grow to giant size.

In the Pacific Northwest, trees were killed by the Mount St. Helens eruption before they were scheduled to be cut down. The timber companies had to spend money salvaging what they could. They also had to replant, to replace the lost trees and to prevent further losses from erosion.

THE COSTS OF CLEANING UP

Cities, states, and individual citizens must pay for cleaning up the streets, homes, and vehicles damaged by volcanic products. In some cities, mud and ash must be hauled away and buried. All these unusual cleanup activities take time, money, and equipment away from routine operations that people generally take for granted.

People have to spend money to rebuild or replace houses and businesses. Stores must have new merchandise. If the people or businesses are insured, the insurance companies pay large sums of money in claims. This means that afterward they raise people's premiums to make up for the payouts.

VOLCANOES AND NATURAL BEAUTY

The "before" and "after" pictures of Mount St. Helens (see chapter 1) show how greatly an eruption can change the landscape. The mountain was once a symbol of nature's beauty. Now its profile is a breathtaking reminder of nature's power. The "after" picture shows something else—that nature rebuilds, even following a major eruption. Spirit Lake still shimmers, and the forests have grown again.

CHAPTER 8

VOLCANIC HAZARDS AND RISKS

A volcanic eruption can be one of the most powerful events in nature. The traditional stories and observations of people around the world testify to that. People sometimes think that death and destruction always follow an eruption. Actually, different types of eruptions have different danger levels. Also, in recent years scientists have been able to give people useful warnings so they can leave the area safely. Even so, some people simply ignore warnings of great danger, as happened at Mount St. Helens. Or they don't believe what the scientists tell them.

A volcano warning is based on several factors. Scientists must know the volcano's present behavior—magma movement, earthquakes, and other physical signs. They must also know the volcano's life history, because it tells the frequency of past eruptions. With this information they can predict the likelihood of an eruption in the near future. Another part of the warning is about the dangers of the eruption. Scientists must also predict the hazards from the various types of activity, such as lava flow or rock blast. Finally, they must translate all of this into risk—how likely a person is to be killed or injured from the eruption.

Scientists have a very good idea what will happen

once an eruption begins. This means they can tell nearby residents and government officials what to expect. But there's also a lot of uncertainty involved in making predictions based on what they know. That means lots of questions can't be answered yes or no.

At Mount St. Helens, for instance, scientists didn't expect the May 18, 1980, eruption to happen that day. They also didn't think the eruption would be so powerful. On the other hand, they weren't entirely surprised, either. If someone had asked on May 17, "Is Mount St. Helens going to erupt tomorrow?" scientists wouldn't have said yes, but they couldn't have said no, either.

VOLCANIC HAZARDS

The hazard from any eruption depends on eruptive force and the type and amount of material produced. For one thing, different materials affect different-sized areas, as Figure 13 shows. Also the amount of material, shape of the land, and wind direction determine where the volcanic materials will go and how much land they will cover. Scientists rank the hazards of various types of eruptions as extreme, high, moderate, and low, as Figure 14 shows.

It's hardest to predict how much damage will be caused by an eruption like that of a Mount St. Helens— pyroclastic flow and rock blasts. But scientists are surer about their predictions for the other types.

VOLCANIC RISKS

The risk from an eruption means the chance that any person in the nearby population will be killed or injured from its eruption. Like the chance of rain tomorrow, the chance that someone will be killed or injured in an eruption can change as events change. Volcanic risks are affected by changes in volcanic activity, in

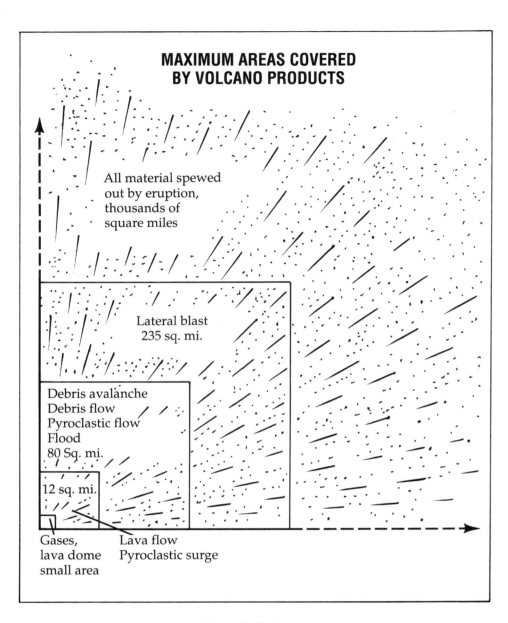

Figure 13. Maximum
areas covered by
volcano products

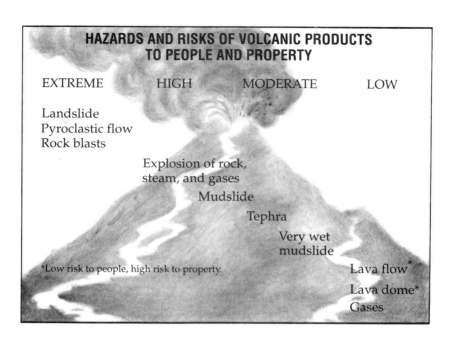

HAZARDS AND RISKS OF VOLCANIC PRODUCTS TO PEOPLE AND PROPERTY

EXTREME   HIGH   MODERATE   LOW

Landslide
Pyroclastic flow
Rock blasts

Explosion of rock,
steam, and gases

Mudslide

Tephra

Very wet
mudslide

*Low risk to people, high risk to property.

Lava flow*

Lava dome*

Gases

human activity, and in knowledge of a volcanic area's history.

CHANGES IN VOLCANIC ACTIVITY

The Long Valley Caldera in northern California's Sierra Nevada Mountains includes the towns of Lee Vining and Bishop. It is the remains of a colossal eruption 730,000 years ago. The 10- by 20-mile area is still very active today, with several craters and domes (Inyo Craters and Domes, and Mono Craters). Scientists aren't sure why the area is still active, but there have been eruptions in the last few hundred years. Since the late 1970s, the Inyo Domes and Mono Craters have had repeated earthquakes.

One scientist thinks the entire area is a grouping of three independent volcanic systems. Each one is oldest

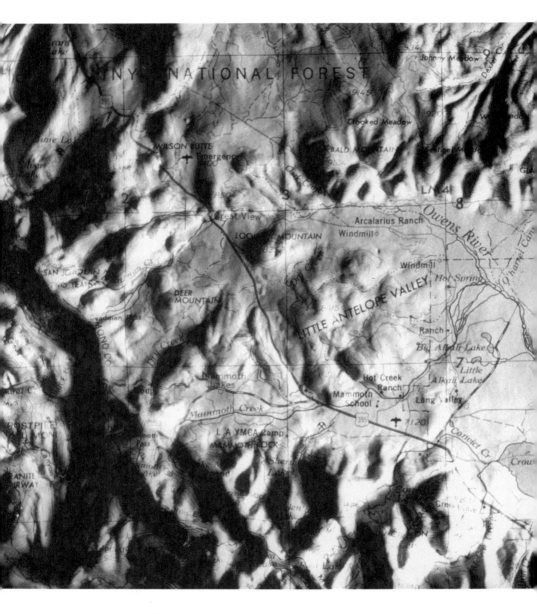

The Long Valley Caldera in northern California is a
high-risk area with a low frequency of volcanic eruptions.

at its south end and youngest and most active at its north end. Do the earthquakes and this theory mean there is a large supply of magma that could trigger an eruption? There's no direct proof, but the many instruments volcano scientists have put there all say that the pool of magma exists.

If the magma is there, where did it come from? Scientists aren't sure. It could be moving up from deep in the mantle. As it moves, some of it could remain in a series of chambers. Pressures from below could heat up the magma and make it active. Then it could cool down.

Is it dangerous? No one can say for certain. About half the time, volcanoes this active simply go back to sleep. The risk to people and property increases as magma becomes more active. Because of the uncertainty about what's happening underground, scientists are observing the area especially closely.

CHANGES IN HUMAN ACTIVITY

Even if a volcanic area's activity doesn't change, an increase in population means an increase in risk. In 1989 and 1990, Redoubt volcano, near Anchorage, Alaska, had several small eruptions of ash and smoke. The last time Redoubt had erupted, in 1967, the metropolitan area was much smaller. As Anchorage and other nearby towns grow in population, the risk from Redoubt grows, too.

CHANGES IN KNOWLEDGE OF A VOLCANO'S HISTORY

Risk can also increase if scientists learn a volcano has been more active than they had thought.

Before the mid-1970s, scientists didn't even know that Mount St. Helens was the youngest and most active volcano in the Cascades. They realized its true risk only when careful field study showed it had erupted thirty-two times during the past 9,000 years, including five times during the nineteenth century. Risk from more remote volcanoes is harder to determine than from volcanoes near large human populations. Many aren't studied until something brings them to people's attention.

For example, Lathrop Wells, an old Nevada volcano, is not on the list of active volcanoes, because scientists thought it hadn't erupted in 200,000 years. The U.S. government wants to store radioactive wastes from nuclear power plants deep inside Yucca Mountain, 12 miles away. These materials are harmful to people's health and the environment, so they must be stored in an escape-proof place. Now scientists have discovered that Lathrop Wells may have erupted only 20,000 years ago, not 200,000.

Does this new information make the radioactive waste storage site more of a risk? Some people think it does. They say Yucca Mountain won't be stable if the nearby volcano erupts. But other people say there's not enough evidence. Lathrop Wells still isn't an active volcano. It's not behaving any differently. But its history has changed.

In this case, the risk can't be measured yet. Lathrop Wells volcano must be studied more closely for clues on how it might behave in the future.

"IS IT GOING TO ERUPT?"

Do you like surprises? Some people don't like surprises at all. They are happier living with routines and regular patterns in their lives. They feel uncomfortable when something unusual happens.

An eruption is unusual. This is a problem, because scientists and government officials have a hard time getting people to take possible eruptions seriously. People usually think that events that don't happen often can't hurt them. In other words, they believe that low frequency equals low risk, even though this isn't always true.

There are different combinations of frequency and risk. For instance, snow avalanches are high frequency/high risk, because they occur frequently in the mountains and can easily kill someone. A gentle spring rain shower happens often in the Southeast, but it isn't very dangerous. It is high frequency/low risk.

A volcanic eruption may be either high or low frequency. This is why it's important to know each volcano's history of eruptions. An eruption may also carry either a high or a low risk. A volcano located in the middle of an unpopulated area may erupt once every hundred years, so one of its eruptions is a low frequency/low risk event. But the eruption of a volcano within a few miles of 10,000 people that erupts once every hundred years is a low frequency/high risk event. It is the hardest kind of event for people to understand.

In Washington State, in the months before the big Mount St. Helens eruption—from March until May 1980—people didn't think the volcano was much of a hazard. This was because an eruption was so rare. Even government officials found it hard to imagine the amount of destruction that an eruption could cause.

In a survey, 50 percent of the residents wanted the volcanic activity to stop. Among tourists, 58 percent wanted there to be a major eruption. As one scientist said, they thought of it as a media event. As such, it couldn't harm them. This kind of thinking is why so many people showed up for the first day of the trout fishing season. It was as if the volcano and its increasing activity didn't exist.

After the eruption, most people in one town 100 miles away said they hadn't known anything might happen. Actually, the town newspaper had printed over sixty articles about the volcano during the three months before the eruption. Many of the stories showed that their town was in the hazardous zone. At least one official even thought there wouldn't be any danger unless lava started flowing. It was a typical reaction to a low frequency/high risk eruption.

By contrast, the people on Hawaii Island live with low-risk eruptions. Kilauea and Mauna Loa erupt gently. Even when lava is flowing, it moves slowly enough to let people leave with their belongings. There's even time to move buildings.

The Hawaii Island volcanoes have different frequency patterns, however. Kilauea has erupted over sixty-five times in the past 140 years. Mauna Loa, which last erupted in 1984, has erupted thirty-five times during the 140 years. Mauna Kea last erupted more than 3,500 years ago and has erupted only twice during the past 5,300 years. Which of the three volcanoes is least likely to erupt? With the frequency patterns, along with measurements of magma activity, the answer is Mauna Kea.

This knowledge and the very low level of seismic activity have permitted the safe construction of several large telescopes on Mauna Kea. Another telescope is being constructed and should begin operating in 1991. If Mauna Kea's eruption frequency were higher, such an observatory might not be built.

LIVING WITH CONTINUING RISK

The people of Long Valley, California, have lived with the possibility of a low frequency/high risk eruption since the late 1970s. How do they get along?

Long Valley is a popular ski area with many resorts. Thousands of visitors come on winter weekends. Residents and officials have had to make several kinds of decisions for their personal safety, the safety of the visitors, and the area's economic health:

How much should we prepare for an eruption?

How much money should we spend on evacuation routes and other emergency procedures?

Should we keep vacation resorts near the active sites from expanding?

Should we publicize the possible risks? Won't that scare skiers and other tourists away?

How much do the scientists really know? Can we believe them? Should we?

Should I move to a safer place?

The people in Long Valley have made tradeoffs in coming to their decisions. They built an evacuation road, but they labeled it a "scenic route," which it is. They permitted some new development near the volcanic areas. They continue to encourage skiers to come to the area. In addition, most people have decided not to move away.

When the volcanic activity increases, people pay more attention to the scientists' warnings. When the activity quiets down, people turn their attention to other things.

Finally, scientists are always careful to say that in half the cases of activity like Long Valley's, nothing happens.

THE NEXT ERUPTION: WHERE AND WHEN?

No one can predict the "next" eruption far in advance, except to say that there will be a next one. Kilauea's continuing eruption has bursts of activity that destroy homes and other buildings as the lava moves toward the sea. The eruption continues into the 1990s.

Volcanic activity in Alaska is likely to continue. Many of the volcanoes there are good candidates for the next eruption.

Will there be an eruption soon in Washington, Oregon, or California? Yes, and there are several good possibilities. Mount Shasta is often mentioned. Mount St. Helens is young and active. It could erupt again. Scientists are watching Long Valley closely. They're also keeping an eye on a less likely prospect, an extremely young volcano called San Francisco Mountain, just above Flagstaff, Arizona. It is showing signs of magma movement, and it is near Sunset Crater, an active volcano that last erupted in 1064.

The unknown is the definition of "soon." Scientists can't predict specific dates, but hope they will be able to in the future.

THE VOLCANIC TREASURE

In India, a legend says that the gods stirred the water around an undersea volcano and unearthed the goddess of wealth, Lakshmi, along with great treasure.

Science has shown that volcanic activity has truly given us treasures—the air we breathe, the land we live on, energy to generate electricity, and beautiful landscapes. Volcanism has left us a record of our planet's history and of our universe. Volcanic events are pipelines from the Earth's dynamic center, letting us see the workings of our great recycling machine.

These days, we might still think of Madame Pele as we wonder at Old Faithful, climb Wizard Island, ski Mammoth Mountain, or watch lava flowing from Kilauea or ash from Redoubt. The power she represents deserves our respect and attention.

GLOSSARY

Aa—Lava that hardens while the flow is moving, forming small chunks with rough, sharp, or jagged surfaces.

Ash—Small pieces of volcanic rock. Hardened lava from earlier eruptions.

Caldera—A large basin-shaped volcanic crater whose diameter is much larger than that of the included vent or vents.

Cinder cone—A cone-shaped mountain built from very small pieces of lava.

Crater—An opening in the Earth caused by a volcanic eruption or by the impact of a meteor.

Crust—The outer layer of Earth rock, covered by sediment or soil.

Fault—A break in the Earth's crust along which there has been movement of one side relative to the other.

Fumarole—A volcanic vent that produces steam or gas.

Geothermal—Referring to heat whose source is in the Earth itself.

Geyser—A fountain of hot water, heated by magma, shot from the Earth with explosive force.

Harmonic tremor—A pattern of Earth movement signaling that an eruption is about to take place.

Igneous—Formed by volcanic action or heat.

Lava—Molten rock on the surface of the Earth.

Lithosphere—The layer of dense rock forming the tectonic plates. See *Plates*.

Magma—Molten rock deep within the Earth. Its heat causes eruptions. If it comes to the surface, it is called lava.

Pahoehoe—Lava that cools on the surface while the flow is moving, forming a smooth, ropy, or pillowy surface texture.

Plates—The large sections that make up the Earth's crust.

Plate tectonics—The generally accepted theory that the Earth's crust is composed of moving plates—part of the system the Earth uses to get rid of heat.

Pyroclastic flow—An eruptive flow of extremely hot rock fragments and volcanic gases.

Shield volcano—A massive volcano built up of layers of lava. Hawaii's volcanoes are examples.

Silica—Silicon dioxide, the main component of lava (and of glass, too).

Solfatara—Volcanic gas made mostly of sulfur.

Strato volcano—A volcano that erupts through a central crater, producing lava sometimes and rocky material at other times. An example is Mount St. Helens.

Volcano—The surface area where magma interacts with the Earth's crust. A variation of the Italian island named Vulcano, which was named after Vulcan, the Roman god of fire.

Volcanology—The scientific study of volcanic activity.

SELECTED BIBLIOGRAPHY

Arizona Bureau of Mines. Geologic Maps, county series. Tucson: University of Arizona, var. dates.

Bercovici, D., G. Schubert, and G.A. Glatzmaier. ''Three-Dimensional Spherical Models of Convection in the Earth's Mantle.'' *Science*, May 26, 1989, p. 950.

Blong, R.J. *Volcanic Hazards. A Sourcebook on the Effects of Eruptions.* Orlando, Fla.: Academic Press, 1984.

Bullard, Fred M. *Volcanoes of the Earth,* 2nd Rev. Ed. Austin: University of Texas Press, 1984.

Bulletin of Volcanology. Global Volcanism Network (formerly SEAN) reports in each issue.

California Division of Mines and Geology. Geologic Data Map Series. Sacramento: var. dates.

Eaton, W.W. *Geothermal Energy.* Washington, D.C.: U.S. Energy Research and Development Administration, 1975.

Foxworthy, B.L., and M. Hill. *Volcanic Eruptions of 1980 at Mt. St. Helens. The First 100 Days.* Geological Survey Professional Paper 1249. Washington, D.C.: U.S. Government Printing Office, 1982.

Lipman, P.W., and D.M. Mullineaux, eds. *The 1980 Eruptions of Mt. St. Helens, Washington.* Geological Survey Professional Paper 1250. Washington, D.C.: U.S. Government Printing Office, 1981.

Macdonald, G.A., and A.T. Abbott. *Volcanoes in the Sea. The Geology of Hawaii,* 2nd ed. Honolulu: University of Hawaii Press, 1987.

Mader, G.G., and M.L. Blair, with VSP Associates, Inc., R.A. Olson. *Living With a Volcanic Threat. Response to Volcanic Hazards, Long Valley, California.* Portola Valley, Calif.: William Spangle and Associates, Inc., 1987.

McClelland, L., et al., eds. *Global Volcanism 1975–1985.* Washington, D.C.: Smithsonian Institution—Prentice-Hall—American Geophysical Union, 1989.

Miller, C.D. *Potential Hazards from Future Volcanic Eruptions in Cali-*

fornia. U.S. Geological Survey Bulletin. 1847. Washington, D.C.: U.S. Government Printing Office, 1989.

Ojakangas, R.W., and D.G. Darby. *The Earth Past and Present.* New York: McGraw-Hill, 1983.

Okal, E.A., ed. *Advances in Volcanic Seismology.* Boston: Birkhauser Verlag, 1987.

Ollier, Cliff. *Volcanoes.* New York: Basil Blackwell, 1988.

Richards, M.A., R.A. Duncan, and V.E. Courtillot. "Flood Basalts and Hot-Spot Tracks: Plumed Heads and Tails." *Science,* October 6, 1989, p. 103.

Rosenfield, Charles, and Robert Cooke. *Earthfire.* Cambridge, Mass.: The MIT Press, 1982.

Simkin, T., et al. *Volcanoes of the World.* Stroudsburg, Pa.: Smithsonian Institution—Hutchinson Ross Publishing Co., 1981.

Solomon, S.C., and J.W. Head. "Mechanisms for Lithospheric Heat Transfer on Venus: Implications for Tectonic Style and Volcanism." *Journal of Geophysical Research,* vol. 87 (B11), 9236, Nov. 11, 1982.

Stewart, J.H., and J.E. Carlson. "Distribution and Lithologic Character of Sedimentary and Igneous Rocks and Unconsolidated Deposits of Nevada Less Than 6 Million Years Old, Showing Centers of Volcanism." Map 52, sheet 4. Carson City, Nevada Bureau of Mines and Geology, 1976.

Tilling, Robert I. *Eruptions of Mount St. Helens: Past, Present, and Future.* Washington, D.C.: U.S. Geological Survey, 1984.

Waring, G.A., R.R. Blankenship, and R. Bentall. *Thermal Springs of the United States and Other Countries of the World—A Summary.* Professional Papers of the U.S. Geological Survey 492, 1-282. Washington, D.C.: U.S. Government Printing Office, 1965.

Weyman, D. *Tectonic Processes.* London: George Allen & Unwin, 1981.

FOR FURTHER READING

Hornby, George. *Your National Parks. A State-by-State Guide.* New York: Crown Publishers, 1980.

Kirk, Ruth. *Exploring Crater Lake Country.* Seattle: University of Washington Press, 1975.

Krafft, Katia. *Volcanoes: Earth's Awakening.* Maplewood, N.J.: Hammond, 1980.

Rosenfeld, Charles, and Robert Cooke. *Earthfire.* Cambridge, Mass.: The MIT Press, 1982.

Rossbacher, Lisa A. *Recent Revolutions in Geology.* New York: Franklin Watts, 1986.

Tatreau, Doug, and Bobbe Tatreau. *Parks of the Pacific Coast. The Complete Guide to the National and Historic Parks of California, Oregon, and Washington.* New York: The East Woods Press, 1985.

Tilling, Robert. *Volcanoes and Igneous Rocks.* New York: Putnam, 1990.

Vogt, Gregory. *Predicting Volcanic Eruptions.* New York: Franklin Watts, 1989.

INDEX